VOLUME ONE HUNDRED AND ONE

ADVANCES IN PARASITOLOGY

Asiatic Liver Fluke - From Basic Science to Public Health, Part A

VOLUME ONE HUNDRED AND ONE

ADVANCES IN PARASITOLOGY

Asiatic Liver Fluke - From Basic Science to Public Health, Part A

Edited by

BANCHOB SRIPA

WHO Collaborating Centre for Research and Control of Opisthorchiasis, Tropical Disease Research Laboratory (Southeast Asian Liver Fluke Disease), Faculty of Medicine, Khon Kaen University, Khon Kaen, Thailand

PAUL J. BRINDLEY

Department of Microbiology, Immunology and Tropical Medicine, and Research Center for Neglected Tropical Diseases of Poverty, School of Medicine & Health Sciences, The George Washington University, Washington DC, United States

ELSEVIER

ACADEMIC PRESS

An imprint of Elsevier

Academic Press is an imprint of Elsevier
125 London Wall, London EC2Y 5AS, United Kingdom
525 B Street, Suite 1650, San Diego, CA 92101, United States
50 Hampshire Street, 5th Floor, Cambridge, MA 02139, United States
The Boulevard, Langford Lane, Kidlington, Oxford OX5 1GB, United Kingdom

First edition 2018

Notices
Knowledge and best practice in this field are constantly changing. As new research and experience broaden our understanding, changes in research methods, professional practices, or medical treatment may become necessary.

Practitioners and researchers must always rely on their own experience and knowledge in evaluating and using any information, methods, compounds, or experiments described herein. In using such information or methods they should be mindful of their own safety and the safety of others, including parties for whom they have a professional responsibility.

To the fullest extent of the law, neither the Publisher nor the authors, contributors, or editors, assume any liability for any injury and/or damage to persons or property as a matter of products liability, negligence or otherwise, or from any use or operation of any methods, products, instructions, or ideas contained in the material herein.

ISBN: 978-0-12-814576-0
ISSN: 0065-308X

Publisher: Zoe Kruze
Acquisition Editor: Ashlie Jackman
Editorial Project Manager: Ana Claudia Garcia
Production Project Manager: Abdulla Sait
Cover Designer: Alan Studholme

Typeset by TNQ Technologies

CONTENTS

CONTRIBUTORS

Ross H. Andrews
Cholangiocarcinoma Screening and Care Program (CASCAP), Faculty of Medicine, Khon Kaen University, Khon Kaen, Thailand; Cholangiocarcinoma Screening and Care Program (CASCAP) and Institute of Cholangiocarcinoma, Khon Kaen University, Khon Kaen, Thailand; Faculty of Medicine, St Mary's Campus, Imperial College, London, United Kingdom

Kanyarat Duenngai
Department of Public Health, Faculty of Science and Technology, Phetchabun Rajabhat University, Phetchabun, Thailand

Steven W. Edwards
Institute of Integrative Biology, University of Liverpool, Liverpool, United Kingdom

Robin B. Gasser
The University of Melbourne, Parkville, VIC, Australia

Melissa Haswell
Faculty of Health, School of Public Health and Social Work, Queensland University of Technology, Brisbane, QLD, Australia

Suwicha Kasemsuwan
Department of Veterinary Public Health, Faculty of Veterinary Medicine, Kasetsart University, Bangkok, Thailand

Nadda Kiatsopit
Department of Parasitology, Faculty of Medicine, Khon Kaen University, Khon Kaen, Thailand

Trevor N. Petney
Cholangiocarcinoma Screening and Care Program (CASCAP), Faculty of Medicine, Khon Kaen University, Khon Kaen, Thailand; Institute of Zoology 1: Ecology and Parasitology, Karlsruhe Institute of Technology, Karlsruhe, Germany; State Museum of Natural History Karlsruhe, Karlsruhe, Germany; Cholangiocarcinoma Screening and Care Program (CASCAP) and Institute of Cholangiocarcinoma, Khon Kaen University, Khon Kaen, Thailand

Waraphon Phimpraphai
Department of Veterinary Public Health, Faculty of Veterinary Medicine, Kasetsart University, Bangkok, Thailand

Sattrachai Prasopdee
Chulabhorn International College of Medicine, Thammasat University, Bangkok, Thailand

Charlotte Price
Institute of Integrative Biology, University of Liverpool, Liverpool, United Kingdom

Prasert Saichua
Tropical Medicine Graduate Program, Faculty of Medicine, Khon Kaen University, Khon Kaen, Thailand; WHO Collaborating Centre for Research and Control of Opisthorchiasis (Southeast Asian Liver Fluke Disease), Department of Pathology, Faculty of Medicine, Khon Kaen University, Khon Kaen, Thailand

Weerachai Saijuntha
Walai Rukhavej Botanical Research Institute, Mahasarakham University, Maha Sarakham, Thailand

Kanin Salao
Department of Microbiology, Faculty of Medicine, Khon Kaen University, Khon Kaen, Thailand

Paiboon Sithithaworn
Cholangiocarcinoma Screening and Care Program (CASCAP), Faculty of Medicine, Khon Kaen University, Khon Kaen, Thailand; Department of Parasitology, Faculty of Medicine, Khon Kaen University, Khon Kaen, Thailand; Cholangiocarcinoma Screening and Care Program (CASCAP) and Institute of Cholangiocarcinoma, Khon Kaen University, Khon Kaen, Thailand

Edward M. Spofford
Institute of Integrative Biology, University of Liverpool, Liverpool, United Kingdom

Banchob Sripa
WHO Collaborating Centre for Research and Control of Opisthorchiasis (Southeast Asian Liver Fluke Disease)/Tropical Disease Research Center (TDRC), Department of Pathology, Faculty of Medicine, Khon Kaen University, Khon Kaen, Thailand; Tropical Medicine Graduate Program, Faculty of Medicine, Khon Kaen University, Khon Kaen, Thailand

Sutas Suttiprapa
Tropical Medicine Graduate Program, Faculty of Medicine, Khon Kaen University, Khon Kaen, Thailand; WHO Collaborating Centre for Research and Control of Opisthorchiasis (Southeast Asian Liver Fluke Disease), Department of Pathology, Faculty of Medicine, Khon Kaen University, Khon Kaen, Thailand

Apiporn Suwannatrai
Department of Parasitology, Faculty of Medicine, Khon Kaen University, Khon Kaen, Thailand

Prasarn Tangkawattana
Faculty of Veterinary Medicine, Khon Kaen University, Khon Kaen, Thailand

Sirikachorn Tangkawattana
Department of Veterinary Pathobiology, Faculty of Veterinary Medicine, Khon Kaen University, Khon Kaen, Thailand; WHO Collaborating Centre for Research and Control of Opisthorchiasis (Southeast Asian Liver Fluke Disease)/Tropical Disease Research Center (TDRC), Department of Pathology, Faculty of Medicine, Khon Kaen University, Khon Kaen, Thailand

Smarn Tesana
Department of Parasitology, Faculty of Medicine, Khon Kaen University, Khon Kaen, Thailand

Helen L. Wright
Institute of Integrative Biology, University of Liverpool, Liverpool, United Kingdom

Neil D. Young
The University of Melbourne, Parkville, VIC, Australia

PREFACE

Banchob Sripa* and Paul J. Brindley[§]

*WHO Collaborating Centre for Research and Control of Opisthorchiasis (Southeast Asian Liver Fluke Disease), Tropical Disease Research Laboratory, Faculty of Medicine, Khon Kaen University, Khon Kaen, Thailand
[§]Department of Microbiology, Immunology and Tropical Medicine, and Research Center for Neglected Tropical Diseases of Poverty, School of Medicine & Health Sciences, The George Washington University, Washington DC, United States

Infections with the fish-borne liver flukes *Opisthorchis viverrini, Opisthorchis felineus* and *Clonorchis sinensis* represent a major public health problem in East Asia and Eurasia, where there are >40 million cases of these trematodiases. *O. viverrini* is endemic in Thailand, Lao PDR, Vietnam and Cambodia (Fedorova et al., 2017; Havelaar et al., 2015; Jongsuksuntigul and Imsomboon, 2003; Ogorodova et al., 2015; Sripa et al., 2007). Humans acquire the infection by eating undercooked fish infected with the metacercariae (MC) of the fluke (Sripa et al., 2011). The larval parasites excyst in the duodenum and migrate into the biliary tree. The parasites mature in about 4 weeks into adult flukes, which graze on biliary epithelia. Fluke eggs are shed into the bile and exit with the fecal stream. When sanitation is less than optimal, these eggs enter freshwater ecosystems where they are ingested by the gastropod snail *Bithynia siamensis*. The parasite develops within the snail, in turn releasing cercariae that seek out and penetrate the skin of a cyprinid fish.

Human infection causes hepatobiliary diseases including cholangitis and periductal fibrosis. In endemic regions, infection is the major risk factor for cholangiocarcinoma (CCA) (Humans, 2012; Mairiang, 2017; Sander and Joung, 2014; Sripa et al., 2007). CCAs are slow-growing adenocarcinomas that metastasize to distant sites because of their proximity to lymphatic vessels. Liver fluke-related CCA is usually only diagnosed at an advanced stage, when the primary cancer is refractory to curative surgical resection (Khuntikeo et al., 2016). The mechanism(s) by which infection initiates transformations that eventually culminate in CCA is not understood, but it is likely to be multifactorial, involving biliary tract and systemic inflammation and associated endogenous nitrosation (Haswell-Elkins et al., 1994; Satarug et al., 1996), secretion of parasite mitogens, other mediators and cofactors including dietary preferences for nitrosamine-rich foods (Blechacz et al., 2011; Bouvard et al., 2009; Brindley, da Costa et al., 2015;

de Martel et al., 2012; Jusakul et al., 2015; Mitacek et al., 1999; Sripa et al., 2007; Sripa et al., 2009; Techasen et al., 2012). Although the prevalence of infection remains high in many regions, along with unacceptably high prevalence of liver fluke infection-induced CCA, 'eco-health' strategies, such as the acclaimed Lawa Lake model (Sripa et al., 2015), are being implemented as a means to unify biological, sociological, educational and public health-based control tactics, including targeted mass chemotherapy with praziquantel, to effectively harness and control the problem.

During recent decades, significant advances have been obtained in the bioscience of the liver fluke-induced disease, including the first report of the genome sequence of *O. viverrini*, advanced understanding of the cellular and molecular biology of this pathogen, advances in disease modelling and epidemiology and immunology and pathogenesis of opisthorchiasis, new diagnostics and treatment as well as a novel control model using integrative ecosystem health approaches to reduce the infection and cancer. This two-volume series will review the state of our knowledge of opisthorchiasis from basic sciences to public health. We anticipate that this volume updates current understanding of *O. viverrini* and opisthorchiasis.

The first volume includes seven chapters, both primary research articles and reviews from leading experts in the field of research on fish-borne liver flukes and opisthorchiasis and neglected tropical diseases. The themes span molecular genetics including the genome and transcriptome, epidemiology, public health and chemotherapy. The authors of Volume 101 Part A are from eight institutes in five countries, including regions where opisthorchiasis is endemic. We hope you will enjoy and be informed by this series that the work and ideas presented advance the field towards better control or even a cure for opisthorchiasis and liver fluke-induced cholangiocarcinoma and that this two-part volume elevates investigation of this problem and pathogen into the more broadly supported research on emerging and infectious diseases.

REFERENCES

Blechacz, B., Komuta, M., Roskams, T., Gores, G.J., 2011. Clinical diagnosis and staging of cholangiocarcinoma. Nat. Rev. Gastroenterol. Hepatol. 8, 512–522.

Bouvard, V., Baan, R., Straif, K., Grosse, Y., Secretan, B., El Ghissassi, F., Benbrahim-Tallaa, L., Guha, N., Freeman, C., Galichet, L., Cogliano, V., Group WHOIAfRoCMW, 2009. A review of human carcinogens–Part B: biological agents. Lancet Oncol. 10, 321–322.

Brindley, P.J., da Costa, J.M., Sripa, B., 2015. Why does infection with some helminths cause cancer? Trends Cancer 1, 174–182.

de Martel, C., Ferlay, J., Franceschi, S., Vignat, J., Bray, F., Forman, D., Plummer, M., 2012. Global burden of cancers attributable to infections in 2008: a review and synthetic analysis. Lancet Oncol. 13, 607–615.

Fedorova, O.S., Kovshirina, Y.V., Kovshirina, A.E., Fedotova, M.M., Deev, I.A., Petrovskiy, F.I., Filimonov, A.V., Dmitrieva, A.I., Kudyakov, L.A., Saltykova, I.V., Odermatt, P., Ogorodova, L.M., 2017. Opisthorchis felineus infection and cholangiocarcinoma in the Russian Federation: a review of medical statistics. Parasitol. Int. 66, 365–371.

Haswell-Elkins, M.R., Satarug, S., Tsuda, M., Mairiang, E., Esumi, H., Sithithaworn, P., Mairiang, P., Saitoh, M., Yongvanit, P., Elkins, D.B., 1994. Liver fluke infection and cholangiocarcinoma: model of endogenous nitric oxide and extragastric nitrosation in human carcinogenesis. Mutat. Res. 305, 241–252.

Havelaar, A.H., Kirk, M.D., Torgerson, P.R., Gibb, H.J., Hald, T., Lake, R.J., Praet, N., Bellinger, D.C., de Silva, N.R., Gargouri, N., Speybroeck, N., Cawthorne, A., Mathers, C., Stein, C., Angulo, F.J., Devleesschauwer, B., World Health Organization Foodborne Disease Burden Epidemiology Reference G, 2015. World Health Organization Global Estimates and Regional Comparisons of the burden of foodborne disease in 2010. PLoS Med. 12, e1001923.

Humans IWGotEoCRt, 2012. Biological agents. Volume 100 B. A review of human carcinogens. IARC Monogr. Eval. Carcinog. Risks Hum. 100, 1–441.

Jongsuksuntigul, P., Imsomboon, T., 2003. Opisthorchiasis control in Thailand. Acta Trop. 88, 229–232.

Jusakul, A., Kongpetch, S., Teh, B.T., 2015. Genetics of *Opisthorchis viverrini*-related cholangiocarcinoma. Curr. Opin. Gastroenterol. 31, 258–263.

Khuntikeo, N., Loilome, W., Thinkhamrop, B., Chamadol, N., Yongvanit, P., 2016. A comprehensive public health conceptual framework and strategy to effectively combat cholangiocarcinoma in Thailand. PLoS Negl. Trop. Dis. 10, e0004293.

Mairiang, E., 2017. Ultrasonographic features of hepatobiliary pathology in opisthorchiasis and opisthorchiasis-associated cholangiocarcinoma. Parasitol. Int. 66, 378–382.

Mitacek, E.J., Brunnemann, K.D., Hoffmann, D., Limsila, T., Suttajit, M., Martin, N., Caplan, L.S., 1999. Volatile nitrosamines and tobacco-specific nitrosamines in the smoke of Thai cigarettes: a risk factor for lung cancer and a suspected risk factor for liver cancer in Thailand. Carcinogenesis 20, 133–137.

Ogorodova, L.M., Fedorova, O.S., Sripa, B., Mordvinov, V.A., Katokhin, A.V., Keiser, J., Odermatt, P., Brindley, P.J., Mayboroda, O.A., Velavan, T.P., Freidin, M.B., Sazonov, A.E., Saltykova, I.V., Pakharukova, M.Y., Kovshirina, Y.V., Kaloulis, K., Krylova, O.Y., Yazdanbakhsh, M., Consortium T, 2015. Opisthorchiasis: an overlooked danger. PLoS Negl. Trop. Dis. 9 e0003563.

Sander, J.D., Joung, J.K., 2014. CRISPR-Cas systems for editing, regulating and targeting genomes. Nat. Biotechnol. 32, 347–355.

Satarug, S., Haswell-Elkins, M.R., Tsuda, M., Mairiang, P., Sithithaworn, P., Mairiang, E., Esumi, H., Sukprasert, S., Yongvanit, P., Elkins, D.B., 1996. Thiocyanate-independent nitrosation in humans with carcinogenic parasite infection. Carcinogenesis 17, 1075–1081.

Sripa, B., Bethony, J.M., Sithithaworn, P., Kaewkes, S., Mairiang, E., Loukas, A., Mulvenna, J., Laha, T., Hotez, P.J., Brindley, P.J., 2011. Opisthorchiasis and Opisthorchis-associated cholangiocarcinoma in Thailand and Laos. Acta Trop. 120 (Suppl. 1), S158–S168.

Sripa, B., Kaewkes, S., Sithithaworn, P., Mairiang, E., Laha, T., Smout, M., Pairojkul, C., Bhudhisawasdi, V., Tesana, S., Thinkamrop, B., Bethony, J.M., Loukas, A., Brindley, P.J., 2007. Liver fluke induces cholangiocarcinoma. PLoS Med. 4, e201.

Sripa, B., Mairiang, E., Thinkhamrop, B., Laha, T., Kaewkes, S., Sithithaworn, P., Tessana, S., Loukas, A., Brindley, P.J., Bethony, J.M., 2009. Advanced periductal fibrosis from infection with the carcinogenic human liver fluke *Opisthorchis viverrini* correlates with elevated levels of interleukin-6. Hepatology 50, 1273–1281.
Sripa, B., Tangkawattana, S., Laha, T., Kaewkes, S., Mallory, F.F., Smith, J.F., Wilcox, B.A., 2015. Toward integrated opisthorchiasis control in northeast Thailand: the Lawa project. Acta Trop. 141, 361–367.
Techasen, A., Loilome, W., Namwat, N., Duenngai, K., Cha'on, U., Thanan, R., Sithithaworn, P., Miwa, M., Yongvanit, P., 2012. *Opisthorchis viverrini*-antigen induces expression of MARCKS during inflammation-associated cholangiocarcinogenesis. Parasitol. Int. 61, 140–144.

CHAPTER ONE

Taxonomy, Ecology and Population Genetics of *Opisthorchis viverrini* and Its Intermediate Hosts

Trevor N. Petney[*,§,¶,1], **Ross H. Andrews**[*,||], **Weerachai Saijuntha**[#], **Smarn Tesana**[**], **Sattrachai Prasopdee**[§§], **Nadda Kiatsopit**[**] and **Paiboon Sithithaworn**[*,**]

[*]Cholangiocarcinoma Screening and Care Program (CASCAP), Faculty of Medicine, Khon Kaen University, Khon Kaen, Thailand
[§]Institute of Zoology 1: Ecology and Parasitology, Karlsruhe Institute of Technology, Karlsruhe, Germany
[¶]State Museum of Natural History Karlsruhe, Karlsruhe, Germany
[||]Faculty of Medicine, Imperial College London, London, United Kingdom
[#]Walai Rukhavej Botanical Research Institute, Mahasarakham University, Maha Sarakham, Thailand
[**]Department of Parasitology, Faculty of Medicine, Khon Kaen University, Khon Kaen, Thailand
[§§]Chulabhorn International College of Medicine, Thammasat University, Bangkok, Thailand
[1]Corresponding author: E-mails: trevor.petney@kit.edu, trevor.petney@t-online.de

Contents

Advances in Parasitology, Volume 101
ISSN 0065-308X
https://doi.org/10.1016/bs.apar.2018.05.001

Abstract

There have been considerable advances in our understanding of the systematics and ecology of *Opisthorchis viverrini*; however, this new knowledge has not only clarified but also complicated the situation. We now know that what was once considered to be a single species is, in fact, a species complex, with the individual species being confined to specific wetland areas. There is also a strong genetic association between the members of the *O. viverrini* species complex and their *Bithynia* snail intermediate hosts. Although this does not negate data collected before the recognition of this situation, it does lead to the caveat that regional and temporal variations in data collected may be related to the species examined. The advances in ecology have generally been spatially limited and have led, in part, to contradictory results that may well be related to nonrecognition of the species studied. It may also be related to natural temporal and spatial variation related, for example, to habitat characteristics. To understand the variation present, it will be necessary to conduct long-term (several years at least) sampling projects after defining the genetic characteristics of *O. viverrini* sensu lato and its *Bithynia* snail intermediate hosts.

1. INTRODUCTION

Cancer is the main cause of mortality in Thailand, with liver cancer, predominantly cholangiocarcinoma (CCA), being the most important type (Bundhamcharoen et al., 2011; Centers of Disease Control, 2014; Khuntikeo et al., 2016). CCA has its highest incidences in continental Southeast Asia, particularly in the northeast and north of Thailand (Bridgewater et al., 2014). *Opisthorchis viverrini* sensu lato (s.l.), a recognized group 1 biological carcinogen (IARC, 2011), is the major risk factor for developing CCA in the countries along the Mekong River. The association between *O. viverrini* s.l. and the development of CCA in humans indicates the importance of understanding the transmission cycle of this parasite. This is necessary to understand how, why and where humans become infected and the control measures that can be used to break the cycle of infection and aid in the elimination of the disease.

The basic epidemiological data for *O. viverrini* s.l. prevalence in humans and the incidence of CCA show that there is considerable local variation within north and northeast Thailand (Sripa and Pairojkul, 2008). This suggests that additional biological factors influence the transmission cycle of the parasite, the parasite's effect on its human hosts and/or the likelihood of humans developing this disease. Thus the eco-epidemiology of *O. viverrini* s.l. infection cannot be considered uniform throughout its area of distribution, but each study area and time of sampling should initially be considered unique to be compared with equivalent data from other sites and dates (Saijuntha et al., 2009). This has proven to be particularly useful in studies on *Schistosoma japonica* for which consistent ecological and associated genetic variation has been found between lake and mountain habitats (Mas-Coma et al., 2014; Peng et al., 2010; Rudge et al., 2009; Xu et al., 2006; Yin et al., 2016). The eco-epidemiology of fascioliasis is also strongly dependent on local ecological conditions (Mas-Coma et al., 2014).

2. SYSTEMATIC STATUS AND POPULATION GENETICS OF *OPISTHORCHIS VIVERRINI* S.L.

Since the previous report on the taxonomy of human liver flukes, including *O. viverrini* s.l. (Kaewkes, 2003), a considerable amount of new information has become available; however, rather than clarifying the situation, this information has tended to complicate it.

The systematic status of the liver flukes *O. viverrini*, *O. felineus* and *Clonorchis sinensis* remains controversial. Current phylogenetic trees provide inconsistent relationships as the trees generated are dependent on the genetic marker used in their methods of construction (Fig. 1). For instance, the ITS1 sequence depicts *O. viverrini* s.l. to be more closely related to *O. felineus* than to *C. sinensis*, whereas sequence analyses of the Pm-int9 gene have revealed that *O. viverrini* s.l. and *C. sinensis* are more closely related to each other than to *O. felineus*. To add to the confusion, it has also been suggested that *O. felineus* is more closely related to *C. sinensis* than to *O. viverrini* s.l. based on ITS2 and mitochondrial DNA/amino acid sequences (Cai et al., 2012; Le et al., 2006; Park, 2007; Saijuntha et al., 2014; Sithithaworn et al., 2012b). Recently the report of an *Opisthorchis* species from Vietnam indicates that it is closely related to *O. viverrini* s.l. (Dao et al., 2016). Although Nawa et al. (2015) considered this was likely *Opisthorchis parageminus*, Dorney et al. (2013) found that the duck specimens were morphologically different from this species. Dao et al. (2017b) found that the duck specimens represent a sister taxon to *O. viverrini* and designated

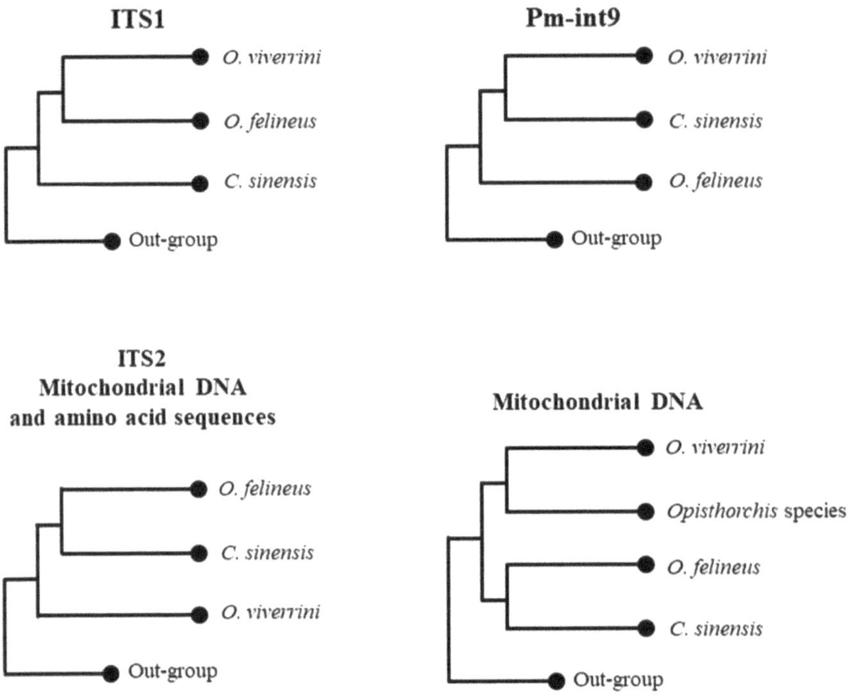

Figure 1 Phylogenetic tree of *Opisthorchis viverrini* s.l. and other sibling species based on different genetic markers analyses.

them as belonging to *Opisthorchis* sp. BD2013. This is of potential significance as, if different species are present, they may differ in their pathological significance to humans (Sithithaworn et al., 2015). Revision at a higher taxonomic level may also be required. Data available in GenBank on nuclear 18S ribosomal DNA and the ribosomal internal transcribed spacer 2 (ITS2) sequences suggest that the families Opisthorchiidae and Heterophyidae cannot be differentiated from one another, with the Opisthorchiidae falling within the Heterophyidae (Thaenkham et al., 2012). The situation is therefore far from clear, and more powerful genetic markers, which have the ability to resolve this controversy, together with increased sample size across a wider geographical area are needed for a comprehensive phylogenetic analysis of these liver flukes (Saijuntha et al., 2014).

Comparison of karyotype formulae has found that *O. viverrini* s.l. is 2n = 12; Lsm4 + Msm2 + Ssm/st2 + Sst/a2 + Sa2 (Kaewkong et al., 2012), whereas *O. felineus* is 2n = 14; Lsm4 + Ssm10 (Polyakov et al., 2010), and *C. sinensis* is either 2n = 14; Lm2 + Lsm2 + 10 small chromosomes or 2n = 56, which is probably an octaploid form of the typical karyotype (Zadesenets et al., 2012).

The molecular systematics and population genetics of *O. viverrini* s.l. has undergone consecutive extensive molecular genetic investigations following the findings by Saijuntha et al. (2007, 2008). Historically, until 2007 there had been few studies on the genetic variation of *O. viverrini* s.l. The first report was published by Sueblinvong et al. (1993). These authors examined the genetic variation in samples of *O. viverrini* s.l. from an autopsy using allozyme markers. Subsequently mitochondrial cytochrome *c* oxidase subunit 1 (cox1) and internal transcribed spacer 2 (ITS2) were used to explore the genetic variation of *O. viverrini* s.l. from different geographical areas in northeast Thailand (Ando et al., 2001). Genetic differentiation between *O. viverrini* s.l. isolates from Thailand and Lao PDR based on RAPD analysis has also been reported (Sithithaworn et al., 2007). These early studies revealed that *O. viverrini* s.l. had varying levels of genetic variation within populations and genetic differentiation between different populations.

Since the first report of liver fluke infection in prisoners in Chiang Mai Province, northern Thailand (Leiper, 1915), only one species, *O. viverrini* s.l., has been recognized by morphology and geographical distribution (Sithithaworn et al., 2012a). Comparison of *O. viverrini* s.l. with the two other related species, *C. sinensis* and *O. felineus*, shows that they can be distinguished by morphological, genetic and geographical differences (Saijuntha et al., 2014).

A comprehensive molecular systematics study by Saijuntha et al. (2007) transformed our perception of the systematic and taxonomic status of *O. viverrini* s.l. by the surprising discovery that it is not a single species but that it is indeed a species complex '*O. viverrini* sensu lato (s.l.)' that contains two evolutionary lineages with many cryptic species (morphologically similar but genetically distinct species) found in distinct wetland systems in Thailand and Lao PDR. Just as importantly, they also discovered evidence of coevolution between *O. viverrini* s.l. and its first intermediate host, *Bithynia* snails, with the location of different genetically distinguishable groups of snails corresponding to that of genetic groups within *O. viverrini* s.l.

Since 2007, RAPD, microsatellite, DNA and RNA molecular genetic studies have supported the existence of these two genetically distinct evolutionary lineages, one found in Thailand and the other in Lao PDR, and have provided more evidence that each lineage contains genetically distinct groups/cryptic species with high genetic variability within and between different populations. These are associated with different geographical areas, temporal factors and fish host species (Ando et al., 2001; Kiatsopit et al., 2011, 2014b; Laoprom et al., 2010, 2012; Sithithaworn et al., 2007). For instance, Laoprom et al. (2010) established and used microsatellite markers to examine the genetic variation and population genetics of *O. viverrini* s.l. and found that a high level of genetic variability occurred within different populations. Subsequent microsatellite analyses showed that the high levels of genetic variability were associated with different geographical areas, temporal factors and/or different species of fish hosts (Pitaksakulrat et al., 2017). Furthermore, an additional potentially cryptic species was detected from Savannakhet Province in Lao PDR by allozyme examination (Kiatsopit et al., 2011). Interestingly, morphologically, genetically and biologically different species of *O. viverrini* s.l. have been discovered in Sakon Nakhon and Nakhon Phanom in northeast Thailand (Laoprom et al., 2009), as well as from Lao PDR (Saijuntha, 2007).

In the Chi River wetland, Thailand, four geographically close populations from Khon Kaen Province were found to be significantly genetically distinct when examined by microsatellite analyses (Laoprom et al., 2012). In the Lao PDR evolutionary lineage, six populations from different localities in Nam Ngum River wetland have been defined into two different genetic groups based on allozyme analyses. These two distinct groups were separated by Nam Ngum Dam (Kiatsopit et al., 2014b). More recent studies using DNA sequence analyses have uncovered a new, genetically distinct cryptic species from the Songkram River wetland, Sakon Nakhon Province, northeast Thailand, which seems to be evolutionarily potentially ancestral to other *O. viverrini* s.l. species (Pitaksakulrat et al., unpublished). There is also a cryptic species from the Sanamchaikate District, Chachoengsao Province, in the central region of Thailand which is found in the Bang Pakong River wetland (Buathong et al., 2017). These findings provide strong evidence that *O. viverrini* s.l. is a complex of cryptic species and genetically distinct groups that form two evolutionary lineages with a wide range of population genetic variation, found in defined wetlands in Thailand and Lao PDR as hypothesized by Saijuntha et al. (2007) (Fig. 2).

Host interactions have also been shown to have an effect on the genetic variation of *O. viverrini* s.l. populations. Multilocus enzyme electrophoresis (MEE) was used to examine the effect of temporal factors and different species of fish host species on the population genetics of *O. viverrini* s.l.

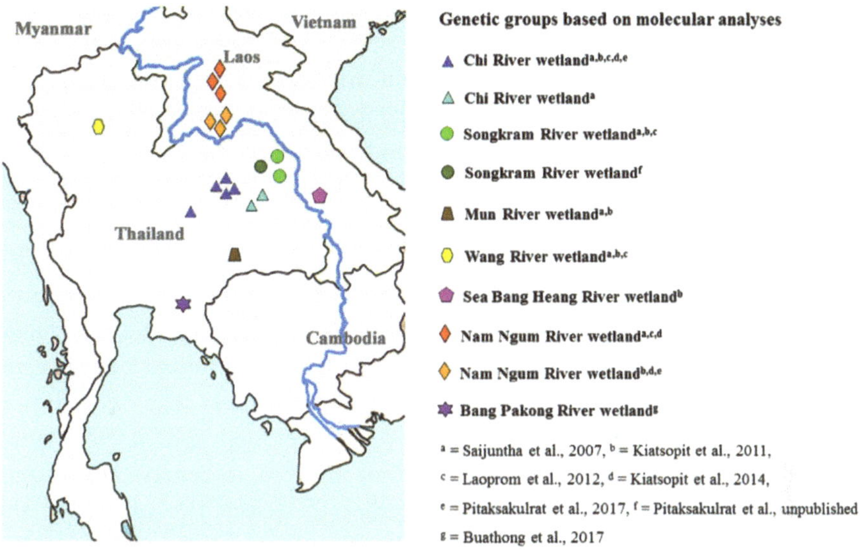

Figure 2 *Opisthorchis viverrini* is a complex of cryptic species and genetically distinct groups in two evolutionary lineages with a wide range of population genetic variation, found in defined wetlands in Thailand and Lao PDR.

from Khon Kaen Province, Thailand (Saijuntha et al., 2009) and Vientiane Province, Lao PDR (Kiatsopit et al., 2014b). These studies have shown that self-fertilization and/or a clonal distribution is commonly found in populations and cryptic species of *O. viverrini* s.l. The role of temporal factors and fish host species appears to have little influence on the levels of genetic differentiation. More recently, in a preliminary evaluation, Pitaksakulrat et al. (2017) used four microsatellite loci for population genetic studies to test the hypothesis that geographical and fish host factors affect the genetic differentiation of *O. viverrini* s.l. in Thailand and Lao PDR. Microsatellite markers distinguished *O. viverrini* s.l. populations into two distinct groups based on geographical areas, which supported previous allozyme studies (Laoprom et al., 2010; Pitaksakulrat et al., 2017). Furthermore, several reports have provided molecular evidence from microsatellite and allozyme analyses suggesting geographical separation as being more important than fish host specificity in determining the genetic structure of *O. viverrini* s. l (Kiatsopit et al., 2014b; Pitaksakulrat et al., 2017; Saijuntha et al., 2009).

Based on the molecular markers that have already been defined and established using different molecular techniques, more comprehensive molecular genetic analyses, for instance, using more defined microsatellite markers, should be used to further explore the population genetics of these *O. viverrini* s.l. populations that occur within the *O. viverrini* s.l. species complex. For more clarity and a better understanding of the molecular systematics of *O. viverrini* s.l., molecular genetic analyses of populations from other mammal hosts, e.g., feline and canine, should be examined to distinguish and define zoonotic and nonzoonotic cryptic species/genetic groups/populations.

A significant avenue for the host selection of *O. viverrini* s.l. is the first intermediate host *Bithynia* snails, which are the critical amplifying stage in the three-host life cycle. This requires more comprehensive molecular analyses of *O. viverrini* s.l. cercariae from these snails in different populations.

Indeed, it is time that the molecular genetic markers that have already been established and those currently being developed and/or fine-tuned from a number of different independent molecular/genetic techniques be used to conduct comprehensive studies to define the extent of the *O. viverrini* s.l. species complex, that is, its systematics, throughout Southeast Asia (Andrews et al., 2008). Investigations of the systematics of medically important trematodes, including the small liver fluke *O. viverrini* s.l., have used both traditional and increasingly sophisticated methodologies to heighten our understanding of host—parasite interactions and disease pathogenesis mechanisms and to enhance our understanding of the carcinogenesis process.

3. THE DISTRIBUTION OF *OPISTHORCHIS VIVERRINI* S.L.

O. viverrini s.l. is only known from continental Southeast Asia, particularly the Mekong Subregion (Sithithaworn et al., 2012a). It is found abundantly in Thailand, mainly in the north and northeast of the country, and Lao PDR, as well as in parts of Cambodia and southern Vietnam, although data for these countries are relatively limited (Doanh and Nawa, 2016; Sithithaworn et al., 2012a; Sohn et al., 2012; Yong et al., 2012). It was also recently reported for the first time from rural populations in Myanmar; however, the extent of its distribution and medical significance in this country remain unknown (Aung et al., 2017). The boundaries of the distribution remain unclear. In Vietnam there is a probable area of overlap between *O. viverrini* s.l. in the south and *C. sinensis* in the north (Doanh and Nawa, 2016), but whether *C. sinensis* replaces *O. viverrini* s.l. in the northeast of Lao PDR along the Vietnamese border is unknown.

Although the prevalence of human infection with *O. viverrini* s.l. in Central and Southern Thailand is lower than that in the north or northeast, some communities are affected (Rangsin et al., 2009; Suwannahitatorn et al., 2013; Traub et al., 2009). This is potentially via immigration from areas where the parasite is more common which tend to be poor agricultural areas (Rigg and Salamanca, 2011). Pumidonming et al. (2018) surveyed villages in nine provinces in lower northern Thailand for human infection with *O. viverrini* s.l. They found prevalences ranging between 2.1% and 28.7% with the majority of infected individuals having migrated from the northeast of Thailand: these individuals had a 2.28- to 2.42-fold higher risk of infection than nonmigrants. Such population movement may also account for the presence of *C. sinensis* in lower Thailand area (Traub et al., 2009).

4. LIFE CYCLE OF *OPISTHORCHIS VIVERRINI*

O. viverrini s.l. has a complex life cycle with freshwater snails belonging to the genus *Bithynia* (*Bithynia funiculata* and two subspecies of *Bithynia siamensis*, *B. s. goniomphalos* and *B. s. siamensis*) as the first intermediate hosts in which it goes through an asexual reproductive cycle (Fig. 3). Cyprinid fish act as the second intermediate hosts in which the infective metacercariae develop. Nowadays, its dominant final host is humans, with domestic cats as potentially important zoonotic reservoir hosts that are infected by the parasite, which serve as a source of infection for humans or another species; dogs are less frequently infected (Aunpromma et al., 2012; Petney et al., 2013).

Adult worms inhabit the small bile ducts of their human or reservoir hosts, but in cases of heavy infection, they may live in larger bile ducts, the gall bladder or pancreatic duct (Tansurat, 1971; Viranuvatti and Stitnimankarn, 1972). They lay about 2000–4200 eggs daily (average 3160) (Wykoff et al., 1965) which pass out of the liver with the bile through the ampulla of Vator which drains into the duodenal lumen during digestion. They are then excreted with the host's faeces. If the eggs reach a suitable freshwater body, they mature within 10–14 days (Chanawong and Waikagul, 1991). Once ingested by *Bithynia* snails, the miracidia enclosed in the eggs hatch in digestive tract where they undergo asexual reproduction. The hatching mechanism involves the action of leucine aminopeptidase, mostly at the operculum–eggshell junction (Tesana et al., 2018). After hatching, the miracidia penetrate the snail tissue and transform to sporocysts. Asexual

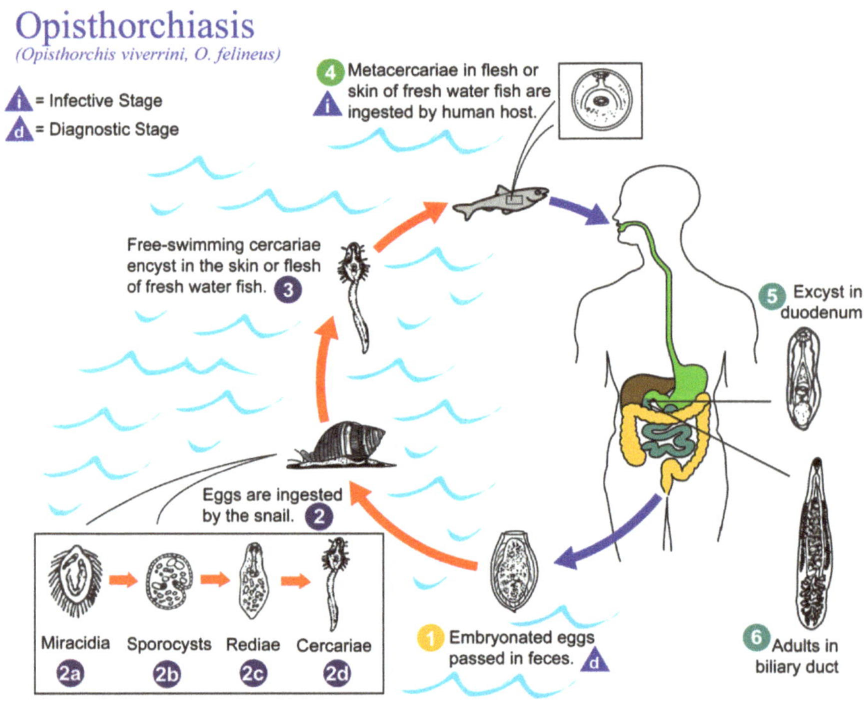

Figure 3 Life cycle of *Opisthorchis viverrini* s.l. (Centers for Disease Control and Prevention image library).

reproduction occurs at the sporocyst and redia stages to produce numerous free-swimming cercariae. The duration from ingestion of the eggs until release of the cercariae takes about 2 months (Wykoff et al., 1965) or from 894 to 951 degree-days (Prasopdee, 2013). *O. viverrini* s.l. cercariae belong to the pleurolophocercous group, with a pair of eyespots and lateral fin folds (Kiatsopit et al., 2016). The fully developed cercariae penetrate through the snail's mantle and are released via the siphon into freshwater.

The cercariae actively swim to find their second intermediate cyprinid fish host (Haas et al., 1990). Once the fish has been found, the cercariae attach to the skin, shedding their tail before penetrating under the skin or into muscle aided by the secretion of a substance from the cystic glands situated laterally on the head (Arunsan et al., 2014). Here they form infective metacercariae. These are oval to spherical in shape and are surrounded by a thick wall produced by the immune response of the fish host (Donthaisong et al., 2014, 2016). They can be distributed throughout the fish body with a high density common in the head (Tesana et al., 1985). Mature metacercaria collected from Khon Kaen Province took about 1 month (443.9–463.7 degree-days) to become infective (Donthaisong et al., 2016).

The definitive host acquires the infection by eating fermented, raw or undercooked freshwater cyprinid fish containing the infective metacercaria. The names of local fish dishes that are the sources of infection include 'larp pla', 'koi pla', 'jom pla noi' and 'pla ra'. The fish protein is digested by pepsin to liberate the metacercariae in the host's stomach. The cyst wall is then digested by trypsin in the duodenum. The larva migrates into the biliary system through the ampulla of Vator, occasionally being found in the pancreatic duct and gall bladder in cases of heavy infection. Worms from Khon Kaen Province were able to produce eggs after about 2 weeks in Syrian hamsters (Khampoosa et al., 2012; Nithikathkul et al., 2007). The entire life cycle of *O. viverrini* s.l. requires 4–4.5 months, and the worms have the lifespan of more than 10 years (Harinasuta and Harinasuta, 1984).

5. ECOLOGICAL CONSIDERATIONS

5.1 Freshwater Ecology in the Mekong Region of Southeast Asia

The ecological background of freshwater systems in the distributional range of *O. viverrini* s.l. provides the framework with which we must consider the ecology of this species complex together with that of its *Bithynia* snail first intermediate and cyprinid fish second intermediate hosts.

5.1.1 Natural Environments

Natural freshwater ecosystems in continental Southeast Asia have extremely high levels of biodiversity (Abell et al., 2008) but have become increasingly rare because of human population pressure and land-use change (Carpenter et al., 2011; Dudgeon, 2000a,b; Hughes, 2017). In addition, they are very poorly understood (Dudgeon, 2003). They therefore currently play very little role in the ecology of *O. viverrini* s.l., although the original discovery of this species in a fishing cat (*Prionailurus viverrinus*, Poirier, 1886) suggests that it was originally a sylvatic species, with a zoonotic life cycle, which switched hosts to humans. Thus the ecological cycle of *O. viverrini* s.l. takes place predominantly in freshwater ecosystems showing a strong human influence.

5.1.2 Human-Modified Environments

There have been massive changes in land use and the human population in continental Southeast Asia over the centuries and most particularly over the last 50–60 years. These include urban and agricultural expansion at the cost of deforestation leading to pollution, overfishing, flow regulation obstructing fish migration, soil degradation and erosion with increased sedimentation and salinization (Allan et al., 2005; Johnson and Kolavalli, 1984; Prachaiyo, 2000; Van Zalinge et al., 2004; Vityakon, 2007), all of which have had a major influence on the freshwater ecology of the region. Included is also the introduction of exotic species such as the tropical catfish *Pterygoplichthys pardalis*, originally from the Amazon basin, as well as the continual, slow change in temperature and rainfall caused by global climate change (Chaichana and Jongphadungkiet, 2012; Dudgeon, 2000a,b).

One of the main land-use changes initiated by humans is the extensive cultivation of rain-fed and irrigated rice (Jongdee et al., 2006; Zeigler and Puckridge, 1995). In association with stocking of the rice paddies with cyprinid second intermediate hosts (Little et al., 1996), such rice cultivation areas provide optimal conditions for the complete life cycle of *O. viverrini* s.l. Kiatsopit et al. (2012) showed that all localities with an exceptionally high prevalence of *B. s. goniomphalos* snails infected with *O. viverrini* s.l. cercariae were rice fields with very shallow water.

In Lao PDR a road improvement scheme aimed to raise the level of the road above the level of flooding caused by the monsoon used aggregates from the properties of local residents with the anticipated advantage of creating ponds for fish production. These were then stocked from local hatcheries with fish, of which a high percentage were already infected with *O. viverrini* s.l. (Sithithaworn et al., 2012a). Given the ability of *Bithynia* snails to move to new areas during the rainy season (Petney et al., 2012) and the frequent lack of sanitary facilities (Erlanger et al., 2008; Opel and Cheuasongkham, 2015), the construction of roads has led to the creation of back-yard ponds in which the life cycle of *O. viverrini* s.l. can be completed (Sithithaworn et al., 2012c).

Ecological studies on *O. viverrini* s.l. to date largely have been of relatively short-term (1 year or less) and/or spatially restricted (one or very few localities). Given the fact that '*O. viverrini* s.l.' contains a number of cryptic species that may be either geographically separated or sympatric and that interannual variation, particularly in climatic factors influencing intermediate hosts, can be substantial, it is difficult to make general ecological conclusions. For any future studies to be of maximum value, they must extend over several years (in the case of global climate change over decades) and include various habitats in geographically well-separated areas. In addition, the flukes found should be genetically defined.

The single, major, long-term study to date (Brockelman et al., 1986) provides a clear indication of the problems involved in short-term studies. *Bithynia* snails were sampled along a series of transects into a natural pond at different depths in Chonnabot District, Khon Kaen Province, Thailand, over a 3-year period from September 1980. This coincided with a 17-month-long drought followed by flooding. During the drought the snails suffered 90% mortality. Samples taken over a single year during this period yielded data that did not indicate the variability present and would have provided a very biased picture of snail and *O. viverrini* s.l. population dynamics. Drought is a frequent problem in the northeast of Thailand (Thavorntam et al., 2015; Wattanakij et al., 2006) and should be taken into consideration, including during studies of *O. viverrini* s.l. in rain-fed rice paddies.

A comparison of four short-term studies on the physical and chemical characteristics of the water in which *B. s. goniomphalos* can be found (Lohachit, 2004–2005; Suwannatrai et al., 2011; Wang et al., 2015; Kim et al., 2016) showed at times substantial differences in temperature, dissolved oxygen, conductivity, salinity and pH, potentially leading to a nonrepresentative subset of the parameter values being considered crucial for the presence of this intermediate host (Wang et al., 2015).

Nevertheless, over the last few years, progress has been made in understanding the ecology of *O. viverrini* s.l. Data indicate that there is considerable temporal and local variation in the dynamics of the snail first intermediate host infection with *O. viverrini* s.l. Although prevalences of infection are generally low, or indeed absent, Kiatsopit et al. (2012) found considerably higher prevalences in some rice field habitats. In one of the most long-term studies to date (Kolporat et al. unpublished data), data from a 4-year period showed substantial, significant interannual and locality-dependent variation in snail size, prevalences of infection with *O. viverrini* s.l. and cercarial release.

There is an increasing tendency to support and carry out more comprehensive studies, including different aspects of the epidemiology of *O. viverrini* s.l. Sripa et al. (2015) have made a preliminary analysis of the limitations of our knowledge of *O. viverrini* s.l. ecology, recommending an ecosystems health approach for determining potential control of opisthorchiasis. Preliminary data from the Lawa project are encouraging, although the authors do point out that more information is required on transmission dynamics and that a continual monitoring of *O. viverrini* s.l. prevalence in all hosts, as well as environmental parameters, is needed. Such data, particularly when available for different areas over several years, should go a long way in filling the gaps we have in our ecological knowledge.

5.2 The Ecology of the *Bithynia* First Intermediate Hosts and Host–Parasite Interactions

5.2.1 Taxonomic Considerations

As early as 2007, Saijuntha et al. (2007) suggested that the genetically characterized populations of *O. viverrini* s.l. corresponded to uniquely genetically characterized populations of their *Bithynia* first intermediate hosts. Kiatsopit et al. (2013) confirmed this. Their results show that two major evolutionary clades characterize *B. s. goniomphalos* with at least nine morphologically similar but genetically distinct cryptic taxa. Clade A is found in the Chi River and the Mun River wetlands in the southern section of northeast Thailand, whereas clade B is found in the Nam Ngum River wetlands in Lao PDR and some localities in the Songkram River wetland in the northern part of northeast Thailand near the Lao border. The status of the *B. s. siamensis* and *B. funiculata* also requires consideration.

As with *O. viverrini* s.l. itself, any study of the *Bithynia* intermediate hosts must define the precise locality where the specimens were collected. Here we use the standard taxonomic designations of *B. funiculata*, *B. s. siamensis* and *B. s. goniomphalos* with the proviso that these are likely to change once detailed taxonomic studies have been carried out.

5.2.2 Distribution

The most well studied first intermediate host is *B. s. goniomphalos* that is distributed from Lao PDR and northeast Thailand to southern Vietnam and Cambodia including the Tonle Sap. *B. s. siamensis* is found in western and central Thailand and in the southern areas of northern Thailand where *B. funiculata* is also found. It has also been found in Mandalay in Myanmar, the Malayan Peninsula and southern areas of Cambodia and southern Vietnam where *B. s. goniomphalso* is also found (Petney et al., 2012).

B. s. goniomphalos is most frequently collected in shallow freshwater environments such as rice fields, roadside ponds and irrigation canals with an underlying red-yellow podzolic soil (Papasarathorn et al., 1990; Suwannatrai et al., 2011; Wang et al., 2015). It can be found on plants in the water as well as mud, sand and solid substrates, such as stones and concrete, of irrigation channels (Papasarathorn et al., 1990; Petney et al., 2012). Wang et al. (2011) suggest that the earthen banks surrounding the rice fields increase water retention time, thus increasing their suitability for the snails. *B. funiculata* also breeds in rice fields (Ngern-klun et al., 2006). Movement from one area to another follows flooding when the snails climb on to floating vegetation that drifts with currents (Petney et al., 2012). The use of rice fields, a human environmental modification, as a dominant habitat for the first intermediate hosts of *O. viverrini* s.l. suggests that major changes in the epidemiological dynamics of this parasite took place with human population growth, leading to the largely anthropogenic final phase of its life cycle (Petney et al., 2013).

5.2.3 The Role of the Physical and Chemical Environment

There are undoubtedly a number of abiotic factors restricting the *Bithynia* species to specific habitats and determining their population density. For instance, *B. s. goniomphalos* is relatively salt tolerant, being found in salinities ranging from 0.05 to 22.11 ppt; however, the population densities in northeast Thailand are highest in salinities ranging from 2.5 to 5.0 ppt and less than 1% soil surface salt (Suwannatrai et al., 2011). As the northeast of Thailand has unusually high soil salt concentrations (Sinanuwong and Takaya, 1974a,b; Suwannatrai et al., 2011), the *Bithynia* hosts from other areas are likely to differ in salt tolerance.

A recent study by Kim et al. (2016, 2017) showed for Lawa Lake in Khon Kaen Province that *B. s. goniomphalos* snail abundance followed a nitrogen-salinity gradient and was highest during the rainy season. In addition, the highest relative abundance of cyprinid fish is found in areas with a high abundance of *Bithynia*. The authors suggest that water contamination with nitrates, which in turn influence algal growth as a food for the snails, provides suitable conditions for the complete *O. viverrini* s.l. life cycle.

Wang et al. (2015) sampled 31 rice fields, streams and ponds that are potential habitats for the snail first intermediate hosts in the Chi watershed in Khon Kaen Province. *B. s. goniomphalos* is found most commonly in rice fields. They found that the habitats differed highly significantly in temperature (higher in rice fields) and pH (lower in rice fields). Differences were also found in dissolved oxygen, conductivity, total dissolved solids and salinity (highest in ponds), but only dissolved oxygen was correlated, negatively, with relative *B. s. goniomphalos* abundance. Overall, temperature, dissolved oxygen, salinity and pH combined accounted for 23.3% of the data variation ($P < .001$) across the entire sampling time with higher values at the beginning of the rainy season (June) and in the cold season (January). Seasonal rice planting and irrigation were correlated with population increases.

Judging by the wide variety of abiotic factors alone that are associated with the presence and abundance of *B. g. goniomphalos*, the population dynamics of species in this complex are under shifting multifactorial control.

Of the species/subspecies of *Bithynia* snails which act as first intermediate hosts of *O. viverrini* s.l., *B. funiculata* is found in the north of Thailand and *B. s. siamensis* in the centre and north (Brandt, 1974; Chitramvong, 1992; Kulsantiwong et al., 2015; Wykoff et al., 1965). *B. funiculata* also prefers shallow water (<30 cm), but conductivity can be as low as 0.00 mS/cm rising up to 0.26 mS/cm, indicating that it prefers lower salinity waters than *B. s. goniomphalos* (Ngern-klun et al., 2006). It can live in relatively turbid water (up to 288 NTU) with a pH range from 6.6 to 7.6.

The process of infection of the snail host is a crucial event that determines the transmission cycle of *O. viverrini* s.l. In endemic areas in northeast Thailand, *O. viverrini* s.l. prevalence in snail hosts is usually very low compared with the infection in fish (more than 90%) and human hosts (Brockelman et al., 1986; Kiatsopit et al., 2012; Prasopdee, 2013; Sri-Aroon et al., 2005; Upatham et al., 1984; Vichasri et al., 1982). Among the *Bithynia* species, *B. funiculata* and *B. s. siamensis* were more susceptible (4—7 times) to *O. viverrini* s.l. infection than *B. s. goniomphalos* under laboratory conditions (Chanawong and Waikagul, 1991). For *B. s. goniomphalos* from Khon Kaen Province, Prasopdee et al. (2015a) showed that the prevalence of infection increased with temperature, reaching a maximum at 34°C, with small specimens (2.0—6.0 mm) being more frequently infected than medium-sized (6.1—10.0 mm) specimens. There was no difference between male and female snails.

Even with the generally low infection rate, the parasites proficiency in using their snail hosts for asexual reproduction ensures high prevalences in their fish hosts (Laoprom et al., 2016, specimens from Sakon Nakhon). This indicates that the snail hosts play a key amplification role in *O. viverrini* s.l. transmission.

It is possible that enhancing the immune response might help eliminate the invading parasite in the snail hosts; thus in addition to diagnosis and treatment of opisthorchiasis in human hosts, a study of snail hosts towards a better understanding of *O. viverrini* s.l.—*Bithynia* interactions is currently attracting significant attention (Prasopdee et al., 2014; Suwannatrai et al., 2016). To provide a molecular infrastructure for research on *O. viverrini* s.l.-*Bithynia* system, the first reference transcriptome of *B. s. goniomphalos* was recently established (Cantacessi et al., 2013). This is a large-scale transcriptomic resource that contributes to the exploration of molecular interactions of snail—parasite interactions. Prasopdee et al. (2014) showed that changes occur in gene expression in the snail upon infection with *O. viverrini* s.l. using high-throughput sequencing. This comparative analysis determined the differences in gene transcription between *O. viverrini* s.l.-infected and *O. viverrini* s.l.-uninfected *B. s. goniomphalos* involving the defence mechanisms of the host. The snails were collected from natural water reservoirs making it impossible to state that the uninfected snails had been never exposed to any parasites or whether they had been previously exposed to *O. viverrini* s.l. and/or other trematodes, which had already been eliminated. However, top upregulated genes in the uninfected snails were found associated with the defence response. Among the upregulated gene transcripts, cathepsin L was strongly involved. This protease is used in breaking down the components of the invading parasites of invertebrates (Myers et al., 2008; Yoshino et al., 1993). Moreover, ficolin- and mucin-encoding genes were also highly expressed. These are known to participate in the innate defence response of invertebrates. Ficolin is believed to play a key role in recognition of pathogen molecules, whereas mucin has been established as a physical barrier to invading pathogens (Hasnain et al., 2013; Zhang and Ali, 2008). The high expression of these genes suggests that the uninfected snails are resistant to or successful in their defence response against *O. viverrini* s.l. infection. While actin- and heat shock protein-encoding genes were largely expressed in infected snails, immune-associated transcripts were not significantly upregulated. The high expression of the actin-encoding gene may be linked to cellular motility and phagocytosis in the cellular-associated immunity of the snail; however, this remains unclear. The role/s of heat shock protein have been shown in involve responses of snail hosts to the invasion of parasites (Ittiprasert et al., 2009; Zahoor et al., 2010). Therefore, the defence response of successfully infected or susceptible snail hosts could be manipulated by *O. viverrini* s.l. infection.

In addition to transcriptomics, the proteomic profile of *B. s. goniomphalos* has been generated (Prasopdee et al., 2015b) showing the sequence of protein generation in experimentally infected snails compared with uninfected individuals. Bierer (2017) studied the biomineralization processes of *B. s. goniomphalos* snail shells for *O. viverrini* s.l.-infected and control snails. The shells were measured and digested, and element concentrations were analyzed. *O. viverrini* s.l.-infected shells were smaller than those that were uninfected. The nickel and zinc concentrations of cercariae-infected snail shells were significantly higher than the concentrations of uninfected shells.

5.2.4 Biotic Relationships

Interactions between *O. viverrini* s.l. and its intermediate hosts and other organisms within the freshwater ecosystem have been poorly studied. A wide variety of trematode species is found within *Bithynia* populations. For example, Kiatsopit et al. (2016) found 20 morphologically distinguishable cercaria types in *B. s. goniomphalos* populations from northeast Thailand and Lao PDR. How these species interact remains to be determined; however, in a metaanalysis Kuris and Lafferty (1994) calculated the expected number of interspecific double infections and found that there were substantially less (a 69% decrease) than there should have been based on a random distribution. They hypothesized that this was due to some form of interspecific competition with subordinate species being eliminated.

Wang et al. (2015) showed a significant negative correlation between the relative abundance of the live-bearing snail *Filopaludina m. martensi* and *B. s. goniomphalos*.

5.2.5 Population Dynamics

Naturally infected *B. s. goniomphalos* from Kalasin Province shed an average of 1728 cercariae/snail daily, with the peak between 08:00 and 10:00 a.m. (Phongsasakulchoti et al., 2005). Specimens from Sakon Nakhon showed peak cercarial emergence over the same period during the hot dry season, but this changed to 10:00 to 12:00 during the rainy cool season (Laoprom et al., 2016). Kiatsopit et al. (2014a) showed similar seasonal variation based on specimens from Lao PDR with peak emergence in the hot dry season from 08:00 to 10:00 and during the cool dry season from 12:00 to 14:00. These data indicate the importance of interseasonal and interregional comparisons. Data from a single area and/or time do not necessarily provide an accurate estimate of what is happening across the distributional area of *O. viverrini* s.l.

5.3 The Ecology of the Cyprinid Second Intermediate Hosts and Host–Parasite Interactions

Substantially less is known about the host–parasite ecology of the cyprinid hosts and *O. viverrini* s.l. than is available for the snails. Although almost no work on the molecular taxonomy of any of the cyprinid second intermediate hosts of *O. viverrini* s.l. has been carried out, there is some evidence of significant genetic variation between areas. Adamson et al. (2009) identified four different genetic groups of *Gymnostomus siamensis* (listed as *Henicorhynchus*, see Kottelat, 2013), the Siamese mud carp known to be infected with *O. viverrini* s.l., in the Mekong River Basin based on mitochondrial DNA. Northern Lao and Thailand populations formed two independent groups that differed from populations from tributaries to the Mekong in Thailand sites above the Khone Falls. A fourth group was defined below the Khone Falls extending to the Mekong Delta. All of these differed significantly from a Klong River population in western Thailand. Hurwood et al. (2008) showed significant local genetic variation in *Gymnostomus lobatus* (listed as *Henicorhynchus,* see Kottelat, 2013) from the Mekong River, two of its tributaries (the Mun and Chi rivers) and the Chao Phraya River. It was hypothesized that the variation originated from historical drainage rearrangements in addition to current ecological differences related to discrete migration pathways. Thus local variation does occur within fish species, and gene flow may be limited between populations from interconnected river systems. Additionally, new and closely related species have been found in a number of cyprinid groups (Chang et al., 2014; Hoang et al., 2015; Kusuma et al., 2016).

In comparison with the first intermediate hosts, there is a much wider host range for the cyprinid hosts, although almost all occur within the family Cyprinidae (Saijuntha et al., 2014; Sithithaworn and Haswell-Elkins, 2003). About 50 species of cyprinid fish from 28 genera have been reported as harbouring *O. viverrini* s.l. metacercariae (Table 1). In addition, *O. viverrini* s.l. has been reported from two noncyprinid genera *Anabas* (Anabantidae) and *Trichopodus* (Osphronemidae) (Ditrich et al., 1990; Komalamisra and Setasuban, 1989; Manivong et al., 2009; Ngoen-klan et al., 2010; Ong et al., 2016; Pinlaor et al., 2013; Rim et al., 2008; Saijuntha et al., 2014; Sithithaworn et al., 1997; Srisawangwong et al., 1997; Sukontason et al., 1999; Touch et al., 2009; Vichasri et al., 1982; Wykoff et al., 1965; Kiatsopit et al., unpublished data). The most common species of cyprinid fish belong to the genera *Cyclocheilichthys, Hampala* and *Puntius* (Pitaksakulrat et al., 2013; Saijuntha et al., 2014). The noncyprinid species *Anabas testudineus* and *Trichopodus microlepis* both have low prevalences and intensities of infection (Eom et al., 2015).

Of the 48 cyprinid species known to harbour *O. viverrini* s.l. metacercariae, 31 have been recorded from Cambodia, Lao PDR, Thailand and Vietnam, 10 from Cambodia, Lao PDR and Thailand, and 9 from a different combination of these countries where *O. viverrini* s.l. is endemic (Table 1). For adjacent countries, 21 species are found in Malaysia, 11 in Myanmar, 14 in China, and 7, mostly introduced, in Singapore (data collated from FishBase in February 2018). It therefore seems likely that the distribution of *O. viverrini* s.l. is not determined by the fish hosts but more likely by the first intermediate *Bithynia* hosts (Petney et al., 2012).

Table 1 A List of Fish Species Known to Harbour *Opisthorchis viverrini* s.l. Metacercaria

Species	*O. viverrini* s.l.-associated Countries	Other Asian Countries (Total Countries Present)
Cyprinidae		
Amblyrhynchichthys truncatus	CLTV	I, Ma (6)
Barbodes rhombeus	CLT	Ma, S (5)
Barbonymus altus	CLTV	(4)
Barbonymus gonionotus	CLTV	I (5); introduced including Ch, Ma, My, P, S, (9)
Barbonymus schwanenfeldii	CLTV	Br, I, Ma, S (8), introduced (2)
Carassius auratus	L; introduced T, V	Ch, My (6); introduced including I, Ma, P, S (78)
Cirrhinus jullieni	CLTV	(4)
Cirrhinus microlepis	CLVT	(4)
Crossocheilus reticulatus	CLTV	Ch (5)
Cyclocheilichthys apogon	CLTV	I, Ma, My (9)
Cyclocheilichthys armatus	CLTV	I, Ma, P (7)
Cyclocheilichthys repasson	CLTV	I, Ma, My (7)
Cyclocheilichthys lagleri	CLT	(3)
Cyclocheilos enoplos	CLTV	I, Ma (6)
Cyclocheilos furcatus	CLT	(3)
Discherodontus ashmeadi	CLT	(3)
Esomus metallicus	CLTV	Ma, My, S (introduced) (7)
Gymnostoma[a] *lineatus*	CLTV	Ch (5)
Gymnostoma lobatus	CLT	(3)
Gymnostoma siamensis	CLTV	(4)
Hampala dispar	CLT	(3)
Hampala macrolepidota	CLTV	Ch, I, Ma, My, S (introduced) (9)
Hypsibarbus lagleri	CLT	(3)
Hypsibarbus pierrei	CLTV	Ch, Ma (7)
Hypsibarbus wetmorei	CLTV	(4)
Labeo chrysophekadion	CLTV	(7)
Labiobarbus leptocheilus	CLVT	Ch, I, Ma, My (8)
Labiobarbus siamensis	CLTV	(4)
Lobocheilus melanotaenia	CLT	(5)
Mystacoleucus atridorsalis	CLT	(3)
Mystacoleucus obtusirostris (syn. *M. marginatus*)	CLTV	Ch, I, Ma, My (8)
Neolissochilus stracheyi	CLTV	(4)

Table 1 A List of Fish Species Known to Harbour *Opisthorchis viverrini* s.l. Metacercaria—cont'd

Species	*O. viverrini* s.l.-associated Countries	Other Asian Countries (Total Countries Present)
Onychostoma elongatum	LV	Ch (3)
Onychostoma fusiforme	L	Ch (2)
Oreichthys parvus	LT	(2)
Osteochilus melanopleurus	CLTV	I, Ma (6)
Osteochilus schlegelii	CTV	I (4)
Osteochilus vittatus	CLTV	Ch, I, Ma, My, S (introduced) (10)
Osteochilus waandersii	CLTV	Br, I, Ma (7)
Paralaubuca barroni	CLT	Ch (4)
Pethia stoliczkana	LT	My (3)
Poropuntius deauratus	V	(1)
Poropuntius laoensis	CLT	(4)
Puntioplites falcifer	CLTV	Ch (5)
Puntioplites proctozysron	CLTV	Ch, Ma (6)
Puntigrus partipentazona	CLTV	Ma, S (introduced) (6)
Puntius brevis	CLTV	I, Ma (6)
Rasbora ourotacniatiran	Not listed in FishBase or Kottelat, 2013	
Systomus rubripinnis (syn. *S. orphoides*)	CLTV	I, Ma, My (8)
Thynnichthys thynnoides	CLTV	Br, I, Ma (7)
Noncyprinids		
Anabas testudineus (Anabantidae)	CLTV	Ch, I, Ma, My, P, S (20)
Channa striata (Channidae)	CLTV	Ch, I, Ma, My, P (23 of which 9 introductions)
Trichopodus microlepis (Osphronemidae)	CTV	S (introduced) (6)

Cambodia (C), Lao PDR (L), Thailand (T) and Vietnam (V) are considered to be the dominant distribution areas of *O. viverrini* s.l. The adjoining Asian countries with known cyprinid hosts are also listed: Brunei (Br), China (Ch), Indonesia (I), Malaysia (Ma), Myanmar (My), Philippines (P), Singapore (S). The species names have been updated based on Kottelat (2013) and countries of distribution are from FishBase 9th February 2018.
[a]*Henicorhynchus* is a junior synonym of *Gymnostoma*.
Modified after Saijuntha, W., Sithithaworn, P., Kaitsopit, N., Andrews, R.H., Petney, T.N., 2014. Liver flukes: Clonorchis and Opisthorchis. Adv. Exp. Med. Biol. 766, 153–199.

The intensity of liver fluke infections in fish varies by season, species, individuals and types of water bodies (Eom et al., 2015; Kaewpitoon et al., 2012; Pinlaor et al., 2013; Sithithaworn et al., 1997; Vichasri et al., 1982). The intensity of infection with metacercariae reported in fish varies within a range of one to thousands (Eom et al., 2015; Rim et al., 2008, 2013). Ngoen-klan et al. (2010) compared the seasonal prevalence and intensity of infection of a number of cyprinid species from sewage treatment wetlands in Cambodia over 2 years. They showed that the fish species caught as well as their relative abundance differed both between years and seasons. Prevalences and intensities of infection also varied, although these tended to be higher in the dry than in the wet season within the same fish species.

Ong et al. (2016) compared the cyprinid fish species assemblages in Ubolratana reservoir with three of its tributary rivers (the results from which were pooled) in Khon Kaen province, northeast Thailand. They found that the rivers were dominated by *Gymnostomus siamensis* (as *Henicorhynchus* in the text) (83.3%). This did not vary substantially between the early rainy, late rainy and cool dry seasons. The abundance of the species in the reservoir was much more even with *Labiobarbus leptocheilus* and *Cyclocheilichthys armatus* being the most common overall making up 27% and 24% of the total catch each, but with variations depending on season. In general, the fish from the reservoir had higher densities (metacercariae/kg) and intensities of infection with *O. viverrini* s.l. indicating that the source of the fish can be of very considerable importance for the transmission cycle. In addition, although not taken into account, there appear to be seasonal differences in the density of infection, indicating that season is also a significant factor in the transmission cycle.

In addition to wild species several species used commonly in aquaculture (the silver barb, *Barbyonemus gonionotus* and the mrigal, *Cirrhinus mrigala*) show moderately high prevalences of infection with *O. viverrini* s.l. (Pitaksakulrat et al., 2013).

The wide variety of fish hosts combined with limited sampling means that the ecology of this part of the life cycle is extremely complex. Nevertheless, this aspect requires urgent attention given that the two sympatric but genetically different *O. viverrini* s.l. populations in Lao PDR are found in different cyprinid species (Pitaksakulrat et al., 2017). If these two populations do represent different species, then the eco-epidemiology of *O. viverrini* s.l. transmission and potentially of CCA epidemiology will depend on which fish species is eaten.

As with the variety of cercariae from different trematode species infecting the *Bithynia* first intermediate hosts, cyprinid fish can harbour metacercariae from a number of different trematode species (Rim et al., 2013; Srisawangwong et al., 1997; Sukontason et al., 1999; Thu et al., 2007). If and how this influences the *O. viverrini* s.l. dynamics remains to be determined.

6. THE ANTHROPOGENIC TRANSMISSION CYCLE: HUMANS TO SNAILS

The dynamics of egg transmission remain obscure, although recent work by Echaubard et al. (2017) have shown that the longevity and infectivity of isolated eggs are temperature dependent, with temperatures above 30°C being detrimental to survival. For temperatures between 25 and 35°C, motile miracidia within the eggs were only found up to 20 days (30°C). At 40°C there was no indication that motile miracidia were present in the experimental eggs. These data indicate that the survival of eggs in the environment is probably relatively short term, weeks rather than months. Based on these data, the authors suggest that the maximum transmission risk is during the rainy season, reaching a peak in September and decreasing during the dry season.

Echaubard et al. (2017) developed methods for infecting *B. s. goniomphalos* under laboratory conditions. After the embryonated eggs are excreted into the environment with the faeces, they must be eaten by an appropriate first intermediate snail host. Here it is likely that the faeces rather than the eggs are attractive as they represent a major potential source of nutrition for the snails, with the amount of energy excreted in human faeces per day being estimated to be approximately 5% of the gross energy intake in adults (Muspratt et al., 2014; Southgate and Durnin, 1970). The snails do not ingest *O. viverrini* s.l. eggs if presented alone but do if they are in faecal material. This is logical as the eggs are of no nutritional value, whereas faeces contain predominantly organic matter with a high nutrition value (Echaubard pers. com.). This means that laboratory experiments can now be used to explore the parasite—snail relationship in detail. Thus hypotheses based on field observations can be tested under controlled conditions.

The major component in human to snail transmission is where and when the snails become infected: this must be near to sources of contaminated human faeces. Kaewkes et al. (2012) showed that the degree of contamination of reservoir or slow flowing water with human faecal bacteria was an indicator of the prevalence of *O. viverrini* s.l. infection in *B. s. goniomphalos*. Wang et al. (2017) found that patches with *B. s. goniomphalos* that had high coliform bacterial contamination were nearer to septic tanks than those that had lower levels of contamination; however, these authors found only a single snail with *O. viverrini* s.l. infection, and it is unclear whether the comparatively large eggs of the fluke can escape from a septic tank. This work can be interpreted as showing that although septic tanks allow bacterial contamination of the environment, they represent a dead end for *O. viverrini* s.l. eggs. Improving hygiene by increasing septic take use should be explored as a means of disrupting the chain of transmission (Petney et al., 2016).

7. THE TRANSMISSION ZOONOTIC CYCLE: CARNIVORES AS RESERVOIR HOSTS

There is still relatively little information on the reservoir hosts of *O. viverrini* s.l. Enes et al. (2010) collected faecal samples from dogs and cats from three villages in an endemic area for *O. viverrini* s.l. in northeast Thailand and found that 36.4% of cats and 3.9% of dogs were infected. Aunpromma et al. (2012), investigating the same area but using a substantially larger sample size, found similar percentages (35.5% and 0.4%, respectively). This pattern seems to be common across the region (Giboda et al., 1991; Scholz et al., 2003).

Intensities of infection were more variable, ranging between 1 and 450 adult worms in cats from Vientiane Province, Lao PDR (Scholz et al., 2003). The intensity of infection, measured as eggs/gram of faeces, was usually higher in cats than in dogs, with the possibility that young dogs are more susceptible than adults (Enes et al., 2010). The average number of eggs/gram recorded by Aunpromma et al. (2012) ranged from 10 to 3692 with a total average of 365, indicating that cats are likely to be significant reservoir hosts. To test this, it would be necessary to characterize cat and human populations of *O. viverrini* s.l. on a molecular basis to determine genetic similarity and gene flow.

Various other families of wild and synanthropic mammals have been examined for internal parasites, including rodents, shrews, cattle, pangolins and primates, but none have been found to harbour *O. viverrini* s. l. (Chaisiri et al., 2010, 2012; Jittapalapong et al., 2010; Mohapatra et al., 2016; Namue and Wongsawad, 1997; Pakdeenarong et al., 2014; Ribas et al., 2016; Veciana et al., 2012; Wenz-Mucke et al., 2013).

8. THE INFLUENCE OF GLOBAL CHANGE

The degree of global change occurring worldwide, particularly in freshwater ecosystems, is massive (Taraschewski and Petney, 2011). There is still considerable uncertainty on how global climate change will influence the hydrology of the Mekong Subregion (Thompson et al., 2014). How global climate change will influence the ecology of *O. viverrini* s.l. and its aquatic intermediate hosts thus remains uncertain. In the Mekong

Subregion, the annual mean temperature is expected to increase associated with a longer annual dry season, more extreme rainfall events and more frequent and severe drought and flooding events (Eastham et al., 2008; Snidvongs, 2006). Based on current estimates, the Mekong River will increase its maximum flow rate and decrease its minimum flow rate (van Vliet et al., 2013). This will potentially increase the risk of flooding in the wet season and drought in the dry season (Cruz et al., 2007; Ziegler, 2012), although extensive dam construction will have a considerable influence on the hydrological patterns (Ziegler et al., 2013). Suwannatrai et al. (2017) used maximum entropy species distribution modelling as a potential predictive model for changes in *O. viverrini* s.l. distribution. They found that rainfall and minimum temperature are currently the determining factors. By 2050, rainfall, maximum temperature and mean temperature during the quarter of maximum rainfall will be significant, and rainfall during the coldest quarter, and maximum and minimum temperatures will be critical by 2070. These predictions suggest that southern and central Thailand will remain climatically largely unsuitable for *O. viverrini* s.l. and that there will be some reduction in suitability in the north and northeast of the country.

Changes in rain-fed rice production with the associated habitat availability for the intermediate hosts will definitely impact *O. viverrini* s.l. ecology. Mainuddin et al. (2013) predict increased rice yield in Thailand and Lao PDR for the period 2010–2050 but a decrease in Cambodia and southern Vietnam. In contrast, based on both climate change and dam construction for hydropower, Arias et al. (2014) predict a substantial increase in rice production in the Tonle Sap area of Cambodia, although this will be associated with a deterioration of the ecological system including a reduction in fish production. As movement of *Bithynia* species occurs as a normal factor during flooding (Petney et al., 2012), the stocking of rice fields with the cyprinid hosts of *O. viverrini* s.l. will lead to increased areas where the life cycle of the parasite can be completed.

9. FUTURE RESEARCH PRIORITIES

First and foremost it is absolutely necessary to determine the taxonomic status of the *O. viverrini* s.l. from different areas and the various intermediate hosts. The eco-epidemiology of *O. viverrini* s.l. is undergoing dynamic changes caused by a variety of predominantly man-made factors. These include global climate change, land-use change, human population growth, as well as preventive measures such as cure using praziquantel, enhanced education (Khuntikeo et al., 2016; Laithavewat et al., 2018), better sanitation (Petney et al., 2016) and screening (Khuntikeo et al., 2016). To observe and understand these dynamic processes, identify their causes and predict their outcomes for the human population, it will be necessary to create a long-term, preferably continuous field research program to determine what changes are occurring in the distributions and abundances of both the first and second intermediate hosts, as well as in parasite intensity and prevalence in all hosts.

O. viverrini infection in humans is usually considered as an individual infection. This is also the case for suggested One-Health systems for addressing opisthorchiasis (Sripa et al., 2015). However, in developing and other tropical countries, infections with multiple parasites species are common, especially in rural areas (Petney and Andrews, 1998). This is true for rural Thailand, Lao PDR and Cambodia in areas where *O. viverrini* s.l. is found (Boonjaraspinyo et al., 2013; Herter et al., 2007; Sayasone et al., 2011; Soukhathammavong et al., 2012; Yong et al., 2012). How do multiple infections influence pathology? Do multimorbidity factors influence the likelihood of developing cholangiocarcinoma?

There are also various inconsistencies between studies that should be further investigated. It is likely that these are not the result of inadequate or varying techniques but represent, at least in part, natural variability due to complex temporal and spatial environmental interactions that are currently inadequately understood or due to differences in the biology of cryptic species within the *O. viverrini* s.l. complex.

There is a great deal of variation in the prevalence of infection in *Bithynia* species. This has been reviewed by Petney et al. (2012) with substantial new information in Kiatsopit et al. (2012). In general, infection rates are very low, frequently being 0% or, when present, 1% or less; however, exceptional cases have been reported with prevalences of up to 8.37% (Kiatsopit et al., 2012).

Some studies have found snail size to be an important factor in infection. These include infection only in reproductive size *B. s. goniomphalos* >8 mm (Brockelman et al., 1986) and in immature *B. s. siamensis* 2–4 mm in size which were more susceptible to infection than mature field caught specimens (6–8 mm) (Chanawong and Waikagul, 1991; Upatham and Sukhapanth, 1980). These differences could be species specific, but Laoprom et al. (2016) found that the relationship between prevalence and snail size in *B. s. goniomphalos* differed between seasons with large snails (>10.0 mm) being more commonly infected in the hot dry season and small-(<8.0 mm) and medium-sized snails in the rainy and cool dry seasons (Laoprom et al., 2016). In Vietnam, there was no correlation between shell size and prevalence of infection for either *B. s. goniomphalos* or *B. s. siamensis* (Dao et al., 2017a).

O. viverrini s.l. is only one of at least 20 species of trematode infecting *B. s. goniomphalos* (Kiatsopit et al., 2016). These authors found it to be only the fourth most common species. A number of questions raised in this publication remain unanswered: How do the trematode species interact within their snail hosts? How often do mixed infections occur? Do different species inhibit infection with other species?

Cercarial shedding was discussed previously. It is clear that there are substantial seasonal and regional differences in the peak time of shedding (Kiatsopit et al., 2016; Laoprom et al., 2016; Phongsasakulchoti et al., 2005), and data from a 4-year study by Kolporat et al. (unpublished data) show that there is also variation with sites and seasons and between years. The factors influencing this remain unknown.

It will also be necessary to investigate the complex cyprinid–parasite ecology. Why do differences in prevalence and intensity of infection occur between species, seasons and sampling sites? This also applies to prevalences and intensities of infection in humans. Such investigations will require a multidisciplinary approach, including molecular studies, field sampling, epidemiology and human geography, as well as patterns of water flow.

ACKNOWLEDGEMENTS

The authors thank the team at CASCAP, Khon Kaen University, for the immense help and support that they have provided over the last 5 years. This work was supported by Khon Kaen University through CASCAP and the National Research Council of Thailand through the Fluke Free Thailand Project. The authors also thank the German Ministry for Education and Research (Bundes Ministerium für Bildung und Forschung; OVISEA project 01DP1006), the National Science Foundation of Thailand and Khon Kaen University for providing funds for workshops at which the themes presented in this article were discussed.

REFERENCES

Abell, R., Thieme, M.L., Revenga, C., Bryer, M., Kottelat, M., Bogutskaya, N., Coad, B., Mandrak, N., Balderas, S.C., Bussing, W., Stiassny, M.L., Skelton, P., Allen, G.R., Unmack, P., Naseka, A., Ng, R., Sindorf, N., Robertson, J., Armijo, E., Higgins, J.V., Heibel, T.J., Wikramanayake, E., Olson, D., López, H.L., Reis, R.E., Lundberg, J.G., Sabaj Pérez, M.H., Petry, P., 2008. Freshwater ecoregions of the world: a new map of biogeographic units for freshwater biodiversity conservation. BioScience 58, 403–414.

Adamson, E.A., Hurwood, D.A., Baker, A.M., Mather, P.B., 2009. Population subdivision in Siamese mud carp *Henicorhynchus siamensis* in the Mekong River basin: implications for management. J. Fish. Biol. 75, 1371–1392.

Allan, J.D., Abell, R., Hogan, Z., Revenga, C., Taylor, B.W., Welcomme, R.L., Winemiller, K., 2005. Overfishing of Inland waters. BioScience 55, 1041–1051.

Ando, K., Sithithaworn, P., Nuchjungreed, C., Tesana, S., Srisawangwong, T., Limviroj, W., Chinzei, Y., 2001. Nucleotide sequence of mitochondrial CO I and ribosomal ITS II genes of *Opisthorchis viverrini* in northeast Thailand. Southeast Asian J. Trop. Med. Publ. Health 32 (Suppl. 2), 17–22.

Andrews, R.H., Sithithaworn, P., Petney, T.N., 2008. *Opisthorchis viverrini*: an underestimated parasite in world health. Trends Parasitol. 24, 497–501.

Arias, M.E., Cochrane, T.A., Kummu, M., Lauri, H., Holtgrieve, G.W., Koponen, J., Piman, T., 2014. Impacts of hydropower and climate change on drivers of ecological productivity of Southeast Asia's most important wetland. Ecol. Model. 272, 252–263.

Arunsan, P., Donthaisong, C., Suwannatrai, K., Prasopdee, S., Kulsantiwong, J., Suwannatrai, A., Boonmars, T., Welbat, J.U., Oopachitakul, S., Tesana, S., 2014. Localization of the cystogenous glands of *Opisthorchis viverrini* cercariae. Southeast Asian J. Trop. Med. Publ. Health 45, 276–289.

Aung, W.P.P., Htoon, T.T., Tin, H.H., Thinn, K.K., Sanpool, O., Jongthawin, J., Sadaow, L., Phosuk, I., Rodpai, R., Intapan, P.M., Maleewong, W., 2017. First report and molecular identification of *Opisthorchis viverrini* infection in human communities from Lower Myanmar. PLoS One 12, e0177130.

Aunpromma, S., Tangkawattana, P., Papirom, P., Kanjampa, P., Tesana, S., Sripa, B., Tangkawattana, S., 2012. High prevalence of *Opisthorchis viverrini* infection in reservoir hosts in four districts of Khon Kaen Province, an opisthorchiasis endemic area of Thailand. Parasitol. Int. 61, 60–64.

Bierer, A., 2017. Influence of Parasitization with *Opisthorchis viverrini* and Xiphidiocercariae on the Biomineralization of Shells from *Bithynia siamensis goniomphalos* (B.Sc. thesis). Institute of Applied Geosciences, Department of Aquatic Geochemistry/Geochemistry and Economic Geology. Karlsruhe Institute of Technology, Germany.

Boonjaraspinyo, S., Boonmars, T., Kaewsamut, B., Ekobol, N., Laummaunwai, P., Aukkanimart, R., Wonkchalee, N., Juasook, A., Sriraj, P., 2013. A cross-sectional study on intestinal parasitic infections in rural communities, northeast Thailand. Kor. J. Parasitol. 51, 727–734.

Brandt, R.A.M., 1974. The non-marine aquatic Mollusca of Thailand. Arch. Molluskenkd. 105, 1–423.

Bridgewater, J., Galle, P.R., Khan, S.A., Llovet, J.M., Park, J.W., Patel, T., Pawlik, T.M., Gores, G.J., 2014. Guidelines for the diagnosis and management of intrahepatic cholangiocarcinoma. J. Hepatol. 60, 1268–1289.

Brockelman, W.Y., Upatham, E.S., Viyanant, V., Ardsungnoen, S., Chantanawat, R., 1986. Field studies on the transmission of the human liver fluke, *Opisthorchis viverrini*, in northeast Thailand: population changes of the snail intermediate host. Int. J. Parasitol. 16, 545–552.

Buathong, S., Leelayoova, S., Mungthin, M., Ruang-Areerate, T., Naaglor, T., Suwannahitatorn, P., Piyaraj, P., Taamasri, P., Tan-Ariya, P., 2017. Molecular discrimination of *Opisthorchis*-like eggs from residents in a rural community of central Thailand. PLoS Neglected Trop. Dis. 11, e0006030.

Bundhamcharoen, K., Odton, P., Phulkerd, S., Tangcharoensathien, V., 2011. Burden of disease in Thailand: changes in health gap between 1999 and 2004. BMC Publ. Health 11, 53.

Cai, X.Q., Liu, G.H., Song, H.Q., Wu, C.Y., Zou, F.C., Yan, H.K., Yuan, Z.G., Lin, R.Q., Zhu, X.Q., 2012. Sequences and gene organization of the mitochondrial genomes of the liver flukes *Opisthorchis viverrini* and *Clonorchis sinensis* (Trematoda). Parasitol. Res. 110, 235–243.

Cantacessi, C., Prasopdee, S., Sotillo, J., Mulvenna, J., Tesana, S., Loukas, A., 2013. Coming out of the shell: building the molecular infrastructure for research on parasite-harbouring snails. PLoS Neglected Trop. Dis. 7, e2284.

Carpenter, S.R., Stanley, E.H., Vander Zanden, M.J., 2011. State of the world's freshwater ecosystems: physical, chemical, and biological changes. Annu. Rev. Environ. Resour. 36, 75–99.

Centers of Disease Control, 2014. CDC in Thailand Fact Sheet. Centers of Disease Control, Atlanta, pp. 1–2.

Chaichana, R., Jongphadungkiet, S., 2012. Assessment of the invasive catfish *Pterygoplichthys pardalis* (Castelnau, 1855) in Thailand: ecological impacts and biological control alternatives. Trop. Zool. 25, 173–182.

Chaisiri, K., Chaeychomsri, W., Siruntawineti, J., Ribas, A., Herbreteau, V., Morand, S., 2012. Diversity of gastrointestinal helminths among murid rodents from northern and northeastern Thailand. Southeast Asian J. Trop. Med. Publ. Health 43, 21–28.

Chaisiri, K., Chaeychomsri, W., Siruntawineti, J., Ribas Salvador, A., Herbreteau, V., Morand, S., 2010. Gastrointestinal helminth infections in Asian house rats (*Rattus tanezumi*) from Northern and Northeastern Thailand. J. Trop. Med. Parasitol 33, 29–35.

Chanawong, A., Waikagul, J., 1991. Laboratory studies on host-parasite relationship of *Bithynia* snails and the liver fluke, *Opisthorchis viverrini*. Southeast Asian J. Trop. Med. Publ. Health 22, 235–239.

Chang, C.H., Li, F., Shao, K.T., Lin, Y.S., Morosawa, T., Kim, S., Koo, H., Kim, W., Lee, J.S., He, S., Smith, C., Reichard, M., Miya, M., Sado, T., Uehara, K., Lavoue, S., Chen, W.J., Mayden, R.L., 2014. Phylogenetic relationships of Acheilognathidae (Cypriniformes: Cyprinoidea) as revealed from evidence of both nuclear and mitochondrial gene sequence variation: evidence for necessary taxonomic revision in the family and the identification of cryptic species. Mol. Phylogenet. Evol. 81, 182–194.

Chitramvong, Y.P., 1992. The Bithyniidae (Gastropoda: Prosobranchia) of Thailand: comparative external morphology. Malacol. Rev. 25, 21–38.

Cruz, R.V., Harasawa, H., Lal, M., Wu, S., Anokhin, Y., Punsalmaa, B., Honda, Y., Jafari, M., Li, C., Huu Ninh, N., 2007. Asia. In: Parry, M.L., Canziani, O.F., Palutikof, J.P., van der Linden, P.J., Hanson, C.E. (Eds.), Climate Change 2007: Impacts, Adaptation and Vulnerability. Contribution of Working Group II to the Fourth Assessment Report of the Intergovernmental Panel on Climate Change. Cambridge University Press, Cambridge, pp. 469–506.

Dao, H.T.T., Dermauw, V., Gabriel, S., Suwannatrai, A., Tesana, S., Nguyen, G.T.T., Dorny, P., 2017a. *Opisthorchis viverrini* infection in the snail and fish intermediate hosts in Central Vietnam. Acta Trop. 170, 120–125.

Dao, T.T., Bui, T.V., Abatih, E.N., Gabriel, S., Nguyen, T.T., Huynh, Q.H., Nguyen, C.V., Dorny, P., 2016. *Opisthorchis viverrini* infections and associated risk factors in a lowland area of Binh Dinh Province, Central Vietnam. Acta Trop. 157, 151–157.

Dao, T.T.H., Nguyen, T.T.G., Gabriel, S., Bui, K.L., Dorny, P., Le, T.H., 2017b. Updated molecular phylogenetic data for *Opisthorchis* spp. (Trematoda: Opisthorchioidea) from ducks in Vietnam. Parasites Vectors 10, 575.

Ditrich, O., Scholz, T., Giboda, M., 1990. Occurrence of some medically important flukes (Trematoda: Opisthorchiidae and Heterophyidae) in Nam Ngum water reservoir, Laos. Southeast Asian J. Trop. Med. Publ. Health 21, 482–488.

Doanh, P.N., Nawa, Y., 2016. *Clonorchis sinensis* and *Opisthorchis* spp. in Vietnam: current status and prospects. Trans. R. Soc. Trop. Med. Hyg. 110, 13–20.

Donthaisong, C., Arunsan, P., Suwannatrai, K., Prasopdee, S., Kulsantiwong, J., Wongmaneeprateep, S., Suwannatrai, A., Tesana, S., 2014. Experimental infection of *Opisthorchis viverrini* cercariae to the cyprinid fish, *Barbonymus gonionotus*. Acta Trop. 136, 118–122.

Donthaisong, C., Arunsan, P., Suwannatrai, K., Prasopdee, S., Kulsantiwong, J., Wongmaneeprateep, S., Suwannatrai, A., Tesana, S., 2016. Infectivity and development of *Opisthorchis viverrini* metacercariae in immunosuppressed *Barbonymus gonionotus* fingerlings (Cyprinidae). Acta Trop. 162, 107–113.

Dorny, P., Dao, T., Victor, B., Nguyen, T., Gabriël, S., 2015. Response to manuscript 'Is *Opisthorchis viverrini* an avian liver fluke?'. J. Helminthol. 89, 257–258.

Dudgeon, D., 2000a. The ecology of tropical Asian rivers and streams in relation to biodiversity conservation. Annu. Rev. Ecol. Systemat. 31, 239–263.

Dudgeon, D., 2000b. Large-scale hydrological changes in tropical Asia: prospects for riverine biodiversity. BioScience 50, 793–806.

Dudgeon, D., 2003. The contribution of scientific information to the conservation and management of freshwater biodiversity in tropical Asia. Hydrobiologia 500, 295–314.

Eastham, J., Mpelasoka, F., Mainuddin, M., Ticehurst, C., Dyce, P., Hodgson, G., Ali, R., Kirby, M., 2008. Mekong River Basin Water Resources Assessment: Impacts of Climate Change. CSIRO, Canberra, 153 pp.

Echaubard, P., Leon, T., Suwanatrai, K., Chaiyos, J., Kim, C.S., Mallory, F.F., Kaewkes, S., Spear, R.C., Sripa, B., 2017. Experimental and modelling investigations of *Opisthorchis viverrini* miracidia transmission over time and across temperatures: implications for control. Int. J. Parasitol. 47, 257–270.

Enes, J.E., Wages, A.J., Malone, J.B., Tesana, S., 2010. Prevalence of *Opisthorchis viverrini* infection in the canine and feline hosts in three villages, Khon Kaen Province, northeastern Thailand. Southeast Asian J. Trop. Med. Publ. Health 41, 36–42.

Eom, K.S., Park, H.S., Lee, D., Sohn, W.M., Yong, T.S., Chai, J.Y., Min, D.Y., Rim, H.J., Insisiengmay, B., Phommasack, B., 2015. Infection status of zoonotic trematode eetacercariae in fishes from Vientiane Municipality and Champasak Province in Lao PDR. Kor. J. Parasitol. 53, 447–453.

Erlanger, T.E., Sayasone, S., Krieger, G.R., Kaul, S., Sananikhom, P., Tanner, M., Odermatt, P., Utzinger, J., 2008. Baseline health situation of communities affected by the Nam Theun 2 hydroelectric project in central Lao PDR and indicators for monitoring. Int. J. Environ. Health Res. 18, 223–242.

Giboda, M., Ditrich, O., Scholz, T., Viengsay, T., Bouaphanh, S., 1991. Current status of food-borne parasitic zoonoses in Laos. Southeast Asian J. Trop. Med. Publ. Health (22 Suppl.), 56–61.

Haas, W., Granzer, M., Brockelman, C.R., 1990. *Opisthorchis viverrini*: Finding and recognition of the fish host by the cercariae. Exp. Parasitol. 71, 422–431.

Harinasuta, C., Harinasuta, T., 1984. *Opisthorchis viverrini*: life cycle, intermediate hosts, transmission to man and geographical distribution in Thailand. Arzneimittelforschung 34, 1164–1167.

Hasnain, S.Z., Gallagher, A.L., Grencis, R.K., Thornton, D.J., 2013. A new role for mucins in immunity: insights from gastrointestinal nematode infection. Int. J. Biochem. Cell Biol. 45, 364–374.

Herter, U., Petney, T., Pipitgool, V., Sithithaworn, P., Vivatpatanakul, K., Hinz, E., Andrews, R., 2007. The influence of pregnancy on intestinal parasite infection in Thai women. Acta Trop. 101, 200–206.

Hoang, H.D., Pham, H.M., Durand, J.D., Tran, N.T., Phan, P.D., 2015. Mahseers genera Tor and Neolissochilus (Teleostei: Cyprinidae) from southern Vietnam. Zootaxa 4006, 551–568.

Hughes, A.C., 2017. Understanding the drivers of Southeast Asian biodiversity loss. Ecosphere 8, e01624.

Hurwood, D.A., Adamson, E.A.S., Mather, P.B., 2008. Evidence for strong genetic structure in a regionally important, highly vagile cyprinid (*Henicorhynchus lobatus*) in the Mekong River Basin. Ecol. Freshw. Fish 17, 273–283.

IARC, 2011. IARC Monographs on the Evaluation of Carcinogenic Risks to Humans Volume 100B Biological Agents: *Opisthorchis viverrini* and *Clonorchis sinensis*. World Health Organization, International Agency for Research on Cancer, Lyon.

Ittiprasert, W., Nene, R., Miller, A., Raghavan, N., Lewis, F., Hodgson, J., Knight, M., 2009. *Schistosoma mansoni* infection of juvenile *Biomphalaria glabrata* induces a differential stress response between resistant and susceptible snails. Exp. Parasitol. 123, 203–211.

Jittapalapong, S., Sangwaranond, A., Nimsuphan, B., Inpankaew, T., Phasufk, C., Pinyopanuwat, N., Chimnoi, W., Kengradomkij, C., Arunwipat, P., Anakewith, T., 2010. Prevalence of gastrointestinal parasitic infections of dairy cows in Thailand. In: Proceedings of the 48th Kasetsart University Annual Conference, Kasetsart University, Bangkok.

Johnson, S.H., Kolavalli, S., 1984. Physical and economic impacts of sedimentation on fishing activities: Nam Pong Basin, Northeast Thailand. Water Int. 9, 185–188.

Jongdee, B., Pantuwan, G., Fukai, S., Fischer, K., 2006. Improving drought tolerance in rainfed lowland rice: an example from Thailand. Agric. Water Manag. 80, 225–240.

Kaewkes, S., 2003. Taxonomy and biology of liver flukes. Acta Trop. 88, 177–186.

Kaewkes, W., Kaewkes, S., Tesana, S., Laha, T., Sripa, B., 2012. Fecal bacterial contamination in natural water reservoirs as an indicator of seasonal infection by *Opisthorchis viverrini* in snail intermediate hosts. Parasitol. Int. 61, 49–51.

Kaewkong, W., Choochote, W., Kanla, P., Maleewong, W., Intapan, P.M., Wongkham, S., Wongkham, C., 2012. Chromosomes and karyotype analysis of a liver fluke, *Opisthorchis viverrini*, by scanning electron microscopy. Parasitol. Int. 61, 504–507.

Kaewpitoon, N., Kaewpitoon, S.J., Ueng-arporn, N., Rujirakul, R., Churproong, S., Matrakool, L., Auiwatanagul, S., Sripa, B., 2012. Carcinogenic human liver fluke: current status of *Opisthorchis viverrini* metacercariae in Nakhon Ratchasima, Thailand. Asian Pac. J. Cancer Prev. 13, 1235–1240.

Khampoosa, P., Jones, M.K., Lovas, E.M., Srisawangwong, T., Laha, T., Piratae, S., Thammasiri, C., Suwannatrai, A., Sripanidkulchai, B., Eursitthichai, V., Tesana, S., 2012. Light and electron microscopy observations of embryogenesis and egg development in the human liver fluke, *Opisthorchis viverrini* (Platyhelminthes, Digenea). Parasitol. Res. 110, 799–808.

Khuntikeo, N., Loilome, W., Thinkhamrop, B., Chamadol, N., Yongvanit, P., 2016. A comprehensive public health Conceptual framework and strategy to effectively combat cholangiocarcinoma in Thailand. PLoS Neglected Trop. Dis. 10, e0004293.

Kiatsopit, N., Sithithaworn, P., Kopolrat, K., Andrews, R.H., Petney, T.N., 2014a. Seasonal cercarial emergence patterns of *Opisthorchis viverrini* infecting *Bithynia siamensis goniomphalos* from Vientiane Province, Lao PDR. Parasites Vectors 7, 551.

Kiatsopit, N., Sithithaworn, P., Kopolrat, K., Namsanor, J., Andrews, R.H., Petney, T.N., 2016. Trematode diversity in the freshwater snail *Bithynia siamensis goniomphalos* sensu lato from Thailand and Lao PDR. J. Helminthol. 90, 312–320.

Kiatsopit, N., Sithithaworn, P., Saijuntha, W., Boonmars, T., Tesana, S., Sithithaworn, J., Petney, T.N., Andrews, R.H., 2012. Exceptionally high prevalence of infection of *Bithynia siamensis goniomphalos* with *Opisthorchis viverrini* cercariae in different wetlands in Thailand and Lao PDR. Am. J. Trop. Med. Hyg. 86, 464–469.

Kiatsopit, N., Sithithaworn, P., Saijuntha, W., Petney, T.N., Andrews, R.H., 2013. *Opisthorchis viverrini*: implications of the systematics of first intermediate hosts, *Bithynia* snail species in Thailand and Lao PDR. Infect. Genet. Evol. 14, 313–319.

Kiatsopit, N., Sithithaworn, P., Saijuntha, W., Pitaksakulrat, O., Petney, T.N., Webster, J.P., Andrews, R.H., 2014b. Analysis of the population genetics of *Opisthorchis viverrini* sensu lato in the Nam Ngum River wetland, Lao PDR, by multilocus enzyme electrophoresis. Parasitol. Res. 113, 2973–2981.

Kiatsopit, N., Sithithaworn, P., Sithithaworn, J., Boonmars, T., Tesana, S., Pitaksakulrat, O., Saijuntha, W., Petney, T.N., Andrews, R.H., 2011. Genetic relationships within the *Opisthorchis viverrini* species complex with specific analysis of *O. viverrini* from Savannakhet, Lao PDR by multilocus enzyme electrophoresis. Parasitol. Res. 108, 211–217.

Kim, C.S., Echaubard, P., Suwannatrai, A., Kaewkes, S., Wilcox, B.A., Sripa, B., 2016. Seasonal and spatial environmental influence on *Opisthorchis viverrini* intermediate hosts, abundance, and distribution: insights on transmission dynamics and sustainable control. PLoS Neglected Trop. Dis. 10, e0005121.

Kim, C.S., Smith, J.F., Suwannatrai, A., Echaubard, P., Wilcox, B., Kaewkes, S., Sithithaworn, P., Sripa, B., 2017. Role of socio-cultural and economic factors in cyprinid fish distribution networks and consumption in Lawa Lake region, Northeast Thailand: novel perspectives on *Opisthorchis viverrini* transmission dynamics. Acta Trop. 170, 85–94.

Komalamisra, C., Setasuban, P., 1989. Heterophyid flukes and *Opisthorchis viverrini*: intensity and rates of infection in cyprinoid fish from an endemic focus in northeast Thailand. J. Trop. Med. Parasitol. 12, 22–28.

Kottelat, M., 2013. The fishes of the Inland waters of Southeast Asia: a catalogue and core bibliography of the fishes known to occur in freshwaters, Mangroves and estuaries. Raffles Bull. Zool. 27, 1–663.

Kulsantiwong, J., Prasopdee, S., Piratae, S., Khampoosa, P., Thammasiri, C., Suwannatrai, A., Boonmars, T., Viyanant, V., Ruangsitichai, J., Tarbsripair, P., Tesana, S., 2015. Trematode infection of freshwater snail, Family Bithyniidae in Thailand. Southeast Asian J. Trop. Med. Publ. Health 46, 396–405.

Kuris, A.M., Lafferty, K.D., 1994. Community structure: larval trematodes in snail hosts. Annu. Rev. Ecol. Systemat. 25, 189–217.

Kusuma, W.E., Ratmuangkhwang, S., Kumazawa, Y., 2016. Molecular phylogeny and historical biogeography of the Indonesian freshwater fish *Rasbora lateristriata* species complex (Actinopterygii: Cyprinidae): cryptic species and west-to-east divergences. Mol. Phylogenet. Evol. 105, 212–223.

Laithavewat, L., Grundy-Warr, C., Khuntikeo, N., Andrews, R.H., Petney, T.N., Yongvanit, P., Banchonhattakit, P., Sithithaworn, P., 2018. The development of a school-based health education model to prevent opisthorchiasis and cholangiocarcinoma in primary school children in northeast Thailand. Glob. Health Promot. in press.

Laoprom, N., Kiatsopit, N., Sithithaworn, P., Kopolrat, K., Namsanor, J., Andrews, R.H., Petney, T.N., 2016. Cercarial emergence patterns for *Opisthorchis viverrini* sensu lato infecting *Bithynia siamensis goniomphalos* from Sakon Nakhon Province, Thailand. Parasitol. Res. 115, 3313–3321.

Laoprom, N., Saijuntha, W., Sithithaworn, P., Wongkham, S., Laha, T., Ando, K., Andrews, R.H., Petney, T.N., 2009. Biological variation within *Opisthorchis viverrini* sensu lato in Thailand and Lao PDR. J. Parasitol. 95, 1307–1313.

Laoprom, N., Sithithaworn, P., Ando, K., Sithithaworn, J., Wongkham, S., Laha, T., Klinbunga, S., Webster, J.P., Andrews, R.H., 2010. Microsatellite loci in the carcinogenic liver fluke, *Opisthorchis viverrini* and their application as population genetic markers. Infect. Genet. Evol. 10, 146–153.

Laoprom, N., Sithithaworn, P., Andrews, R.H., Ando, K., Laha, T., Klinbunga, S., Webster, J.P., Petney, T.N., 2012. Population genetic structuring in *Opisthorchis viverrini* over various spatial scales in Thailand and Lao PDR. PLoS Neglected Trop. Dis. 6, e1906.

Le, T.H., Van De, N., Blair, D., Sithithaworn, P., McManus, D.P., 2006. *Clonorchis sinensis* and *Opisthorchis viverrini*: development of a mitochondrial-based multiplex PCR for their identification and discrimination. Exp. Parasitol. 112, 109–114.

Leiper, R.T., 1915. Notes of the occurrence of parasites, presumably rare, in man. J. Roy. Army Med. Corps 24, 569–575.

Little, D.C., Surintaraseree, P., Innes-Taylor, N., 1996. Fish culture in rainfed rice fields of northeast Thailand. Aquaculture 140, 295–321.

Lohachit, C., 2004–2005. Ecological studies of *Bithynia siamensis goniomphalos*, a snail intermediate host of *Opisthorchis viverrini*, in Khon Kaen province, northeast Thailand. Malacol. Rev. 37/38, 1–26.

Mainuddin, M., Kirby, M., Hoanh, C.T., 2013. Impact of climate change on rainfed rice and options for adaptation in the lower Mekong Basin. Nat. Hazards 66, 905–938.

Manivong, K., Komalamisra, C., Waikagul, J., Radomyos, P., 2009. *Opisthorchis viverrini* metacercariae in cyprinoid fish from three rivers in Khammouane Province, Lao PDR. J. Trop. Med. Parasitol 32, 23–29.

Mas-Coma, S., Valero, M.A., Bargues, M.D., 2014. Fascioliasis. In: Toledo, R., Fried, B. (Eds.), Digenetic Trematodes. Springer, New York, pp. 77–114.

Mohapatra, R.K., Panda, S., Nair, M.V., Acharjyo, L.N., 2016. Check list of parasites and bacteria recorded from pangolins (*Manis* sp.). J. Parasit. Dis. 40, 1109–1115.

Muspratt, A.M., Nakato, T., Niwagaba, C., Dione, H., Kang, J., Stupin, L., Regulinski, J., Mbéguéré, M., Strande, L., 2014. Fuel potential of faecal sludge: calorific value results from Uganda, Ghana and Senegal. J. Water Sanit. Hyg. Dev. 4, 223–230.

Myers, J., Ittiprasert, W., Raghavan, N., Miller, A., Knight, M., 2008. Differences in cysteine protease activity in *Schistosoma mansoni*-resistant and -susceptible *Biomphalaria glabrata* and characterization of the hepatopancreas cathepsin B Full-length cDNA. J. Parasitol. 94, 659–668.

Namue, C., Wongsawad, C., 1997. A survey of helminth infection in rats (*Rattus* spp.) from Chiang Mai Moat. Southeast Asian J. Trop. Med. Publ. Health 28 (Suppl. 1), 179–183.

Nawa, Y., Doanh, P.N., Thaenkham, U., 2015. Is *Opisthorchis viverrini* an avian liver fluke? J. Helminthol. 89, 255–256.

Ngern-klun, R., Sukontason, K.L., Tesana, S., Sripakdee, D., Irvine, K.N., Sukontason, K., 2006. Field investigation of *Bithynia funiculata*, intermediate host of *Opisthorchis viverrini* in northern Thailand. Southeast Asian J. Trop. Med. Publ. Health 37, 662–672.

Ngoen-klan, R., Piangjai, S., Somwang, P., Moophayak, K., Sukontason, K., Sukontason, K.L., Sampson, M., Irvine, K., 2010. Emerging helminths infection in snails and cyprinoid fish in sewage treatment wetlands waters in Cambodia. Asian J. Water Environ. Pollut. 7, 13–21.

Nithikathkul, C., Tesana, S., Sithithaworn, P., Balakanich, S., 2007. Early stage biliary and intrahepatic migration of *Opisthorchis viverrini* in the golden hamster. J. Helminthol. 81, 39–41.

Ong, X., Wang, Y., Sithithaworn, P., Grundy-Warr, C., Pitaksakulrat, O., 2016. Dam influences on liver fluke transmission: fish infection and human fish consumption behavior. Ann. Assoc. Am. Geogr. 106, 755–772.

Opel, A., Cheuasongkham, P., 2015. Faecal sludge management services in rural Lao PDR: critical gaps and important ways forward. In: 3rd Faecal Sludge Management Conference, Hanoi, Vietnam.

Pakdeenarong, N., Siribat, P., Chaisiri, K., Douangboupha, B., Ribas, A., Chaval, Y., Herbreteau, V., Morand, S., 2014. Helminth communities in murid rodents from southern and northern localities in Lao PDR: the role of habitat and season. J. Helminthol. 88, 302–309.

Papasarathorn, T., Sumethanurugkul, P., Viboolyavatana, J., Temcharoen, P., Tongkoom, B., 1990. Studies of life history and ecological requirements of *Bithynia* (Digoniostoma) *siamensis goniomphalos*, the snail intermediate host of *Opisthorchis viverrini*. J. Parasitol. Trop. Med. Assoc. Thailand 3, 21–27.

Park, G.M., 2007. Genetic comparison of liver flukes, *Clonorchis sinensis* and *Opisthorchis viverrini*, based on rDNA and mtDNA gene sequences. Parasitol. Res. 100, 351–357.
Peng, W.X., Tao, B., Clements, A., Jiang, Q.L., Zhang, Z.J., Zhou, Y.B., Jiang, Q.W., 2010. Identifying high-risk areas of schistosomiasis and associated risk factors in the Poyang Lake region, China. Parasitology 137, 1099–1107.
Petney, T.N., Andrews, R.H., 1998. Multiparasite communities in animals and humans: frequency, structure and pathogenic significance. Int. J. Parasitol. 28, 377–393.
Petney, T.N., Andrews, R.H., Saijuntha, W., Wenz-Mucke, A., Sithithaworn, P., 2013. The zoonotic, fish-borne liver flukes *Clonorchis sinensis, Opisthorchis felineus* and *Opisthorchis viverrini*. Int. J. Parasitol. 43, 1031–1046.
Petney, T.N., Lehn, H., Khuntikeo, N., Dolos, K., Norra, S., Holbach, A., Sithithaworn, P., Andrews, R., Pfäffle, M., 2016. Northeast Thailand: New approaches for avoiding water related liver cancer. Water Solutions 157, 118–123.
Petney, T.N., Sithithaworn, P., Andrews, R., Kiatsopit, N., Tesana, S., Grundy-Warr, C., Ziegler, A., 2012. The ecology of the *Bithynia* first intermediate hosts of *Opisthorchis viverrini*. Parasitol. Int. 61, 38–45.
Phongsasakulchoti, P., Sri-aroon, P., Kerdpuech, Y., 2005. Emergence of *Opisthorchis viverrini* cercariae from naturally infected *Bithynia* (Digoniostoma) siamensis goniomphalos. Southeast Asian J. Trop. Med. Publ. Health 36 (Suppl. 4), 189–191.
Pinlaor, S., Onsurathum, S., Boonmars, T., Pinlaor, P., Hongsrichan, N., Chaidee, A., Haonon, O., Limviroj, W., Tesana, S., Kaewkes, S., Sithithaworn, P., 2013. Distribution and abundance of *Opisthorchis viverrini* metacercariae in cyprinid fish in Northeastern Thailand. Kor. J. Parasitol. 51, 703–710.
Pitaksakulrat, O., Kiatsopit, N., Laoprom, N., Webster, B.L., Webster, J.P., Lamberton, P.H.L., Laha, T., Andrews, R.H., Petney, T.N., Blair, D., Carlton, E.J., Spear, R.C., Sithithaworn, P., 2017. Preliminary genetic evidence of two different populations of *Opisthorchis viverrini* in Lao PDR. Parasitol. Res. 116, 1247–1256.
Pitaksakulrat, O., Sithithaworn, P., Laoprom, N., Laha, T., Petney, T.N., Andrews, R.H., 2013. A cross-sectional study on the potential transmission of the carcinogenic liver fluke *Opisthorchis viverrini* and other fishborne zoonotic trematodes by aquaculture fish. Foodb. Pathog. Dis. 10, 35–41.
Poirier, J., 1886. Trematodes nouveaux ou peu connus. Bull. Soc. Philom. Paris 7, 20–40.
Polyakov, A.V., Katokhin, A.V., Bocharova, T.A., Romanov, K.V., L'vova, M.N., Bonina, O.M., Yurlova, M.I., Mordvinov, V.A., 2010. Comparative analysis of karyotypes of *Opisthorchis felineus* from west Siberia. Contemp. Probl. Ecol 3, 1–3.
Prachaiyo, B., 2000. Farmers and forests: a changing phase in Northeast Thailand. Japanese J. Southeast Asian Stud. 38, 271–446.
Prasopdee, S., 2013. Differential Protein Expression of *Bithynia siamensis goniomphalos* Snail upon Infection with Liver Fluke, Opisthorchis viverrini, Parasitology. Ph.D Thesis. Khon Kaen University, Thailand, Khon Kaen University, p. 116.
Prasopdee, S., Kulsantiwong, J., Piratae, S., Khampoosa, P., Thammasiri, C., Suwannatrai, A., Laha, T., Grams, R., Loukas, A., Tesana, S., 2015a. Temperature dependence of *Opisthorchis viverrini* infection in first intermediate host snail, *Bithynia siamensis goniomphalos*. Acta Trop. 141, 112–117.
Prasopdee, S., Sotillo, J., Tesana, S., Laha, T., Kulsantiwong, J., Nolan, M.J., Loukas, A., Cantacessi, C., 2014. RNA-Seq reveals infection-induced gene expression changes in the snail intermediate host of the carcinogenic liver fluke, *Opisthorchis viverrini*. PLoS Neglected Trop. Dis. 8, e2765.
Prasopdee, S., Tesana, S., Cantacessi, C., Laha, T., Mulvenna, J., Grams, R., Loukas, A., Sotillo, J., 2015b. Proteomic profile of *Bithynia siamensis goniomphalos* snails upon infection with the carcinogenic liver fluke *Opisthorchis viverrini*. J. Proteomics 113, 281–291.

Pumidonming, W., Katahira, H., Igarashi, M., Salman, D., Abdelbaset, A.E., Sangkaeo, K., 2018. Potential risk of a liver fluke *Opisthorchis viverrini* infection brought by immigrants from prevalent areas: a case study in the lower Northern Thailand. Acta Trop. 178, 213–218.

Rangsin, R., Mungthin, M., Taamasri, P., Mongklon, S., Aimpun, P., Naaglor, T., Leelayoova, S., 2009. Incidence and risk factors of *Opisthorchis viverrini* infections in a rural community in Thailand. Am. J. Trop. Med. Hyg. 81, 152–155.

Ribas, A., Saijuntha, W., Agatsuma, T., Thongjun, C., Lamsan, K., Poonlaphdecha, S., 2016. Helminths in rodents from wet markets in Thailand. Helminthologia 53, 326–330.

Rigg, J., Salamanca, A., 2011. Connecting lives, living, and location: mobility and spatial signatures in Northeast Thailand, 1982–2009. Crit. Asian Stud. 43, 551–575.

Rim, H.J., Sohn, W.M., Yong, T.S., Eom, K.S., Chai, J.Y., Min, D.Y., Lee, S.H., Hoang, E.H., Phommasack, B., Insisengmay, S., 2008. Fishborne trematode metacercariae detected in freshwater fish from Vientiane Municipality and Savannakhet Province, Lao PDR. Kor. J. Parasitol. 46, 253–260.

Rim, H.J., Sohn, W.M., Yong, T.S., Eom, K.S., Chai, J.Y., Min, D.Y., Lee, S.H., Hoang, E.H., Phommasack, B., Insisiengmay, S., 2013. Fishborne trematode metacercariae in Luang Prabang, Khammouane, and Saravane province, Lao PDR. Kor. J. Parasitol. 51, 107–114.

Rudge, J.W., Lu, D.B., Fang, G.R., Wang, T.P., Basanez, M.G., Webster, J.P., 2009. Parasite genetic differentiation by habitat type and host species: molecular epidemiology of *Schistosoma japonicum* in hilly and marshland areas of Anhui Province, China. Mol. Ecol. 18, 2134–2147.

Saijuntha, W., 2007. A Comprehensive Investigation of Genetic Variation Among Natural Populations of *Opisthorchis viverrini* in Thailand and Lao PDR (Ph.D Thesis). Khon Kaen University, Khon Kaen.

Saijuntha, W., Sithithaworn, P., Chilton, N.B., Petney, T.N., Klinbunga, S., Satrawaha, R., Webster, J.P., Andrews, R.H., 2009. Impact of temporal changes and host factors on the genetic structure of a population of *Opisthorchis viverrini* sensu lato in Khon Kaen Province (Thailand). Parasitology 136, 1057–1063.

Saijuntha, W., Sithithaworn, P., Kaitsopit, N., Andrews, R.H., Petney, T.N., 2014. Liver flukes: *Clonorchis* and *Opisthorchis*. Adv. Exp. Med. Biol. 766, 153–199.

Saijuntha, W., Sithithaworn, P., Wongkham, S., Laha, T., Chilton, N.B., Petney, T.N., Barton, M., Andrews, R.H., 2008. Mitochondrial DNA sequence variation among geographical isolates of *Opisthorchis viverrini* in Thailand and Lao PDR, and phylogenetic relationships with other trematodes. Parasitology 135, 1479–1486.

Saijuntha, W., Sithithaworn, P., Wongkham, S., Laha, T., Pipitgool, V., Tesana, S., Chilton, N.B., Petney, T.N., Andrews, R.H., 2007. Evidence of a species complex within the food-borne trematode *Opisthorchis viverrini* and possible co-evolution with their first intermediate hosts. Int. J. Parasitol. 37, 695–703.

Sayasone, S., Mak, T.K., Vanmany, M., Rasphone, O., Vounatsou, P., Utzinger, J., Akkhavong, K., Odermatt, P., 2011. Helminth and intestinal protozoa infections, multiparasitism and risk factors in Champasack province, Lao People's Democratic Republic. PLoS Neglected Trop. Dis. 5, e1037.

Scholz, T., Uhlirova, M., Ditrich, O., 2003. Helminth parasites of cats from the Vientiane province, Laos, as indicators of the occurrence of causative agents of human parasitoses. Parasite 10, 343–350.

Sinanuwong, S., Takaya, Y., 1974a. Distribution of saline soils in the Khorat Basin of Thailand: preliminary findings. South East Asian Stud. 12, 365–382.

Sinanuwong, S., Takaya, Y., 1974b. Saline soils in Northeast Thailand : Their possible origin as deduced from field evidence. South East Asian Stud. 12, 105–120.

Sithithaworn, P., Andrews, R.H., Nguyen, V.D., Wongsaroj, T., Sinuon, M., Odermatt, P., Nawa, Y., Liang, S., Brindley, P.J., Sripa, B., 2012a. The current status of opisthorchiasis and clonorchiasis in the Mekong Basin. Parasitol. Int. 61, 10–16.

Sithithaworn, P., Andrews, R.H., Petney, T.N., Saijuntha, W., Laoprom, N., 2012b. The systematics and population genetics of *Opisthorchis viverrini* sensu lato: implications in parasite epidemiology and bile duct cancer. Parasitol. Int. 61, 32–37.

Sithithaworn, P., Haswell-Elkins, M., 2003. Epidemiology of *Opisthorchis viverrini*. Acta Trop. 88, 187–194.

Sithithaworn, P., Nuchjungreed, C., Srisawangwong, T., Ando, K., Petney, T.N., Chilton, N.B., Andrews, R.H., 2007. Genetic variation in *Opisthorchis viverrini* (Trematoda: Opisthorchiidae) from northeast Thailand and Laos PDR based on random amplified polymorphic DNA analyses. Parasitol. Res. 100, 613–617.

Sithithaworn, P., Petney, T.N., Andrews, R.H., 2015. What significance do helminths species-complexes have for the prevention, diagnosis and treatment of human infections? Trans. R. Soc. Trop. Med. Hyg. 109, 289–290.

Sithithaworn, P., Pipitgool, V., Srisawangwong, T., Elkins, D.B., Haswell-Elkins, M.R., 1997. Seasonal variation of *Opisthorchis viverrini* infection in cyprinoid fish in north-east Thailand: implications for parasite control and food safety. Bull. World Health Organ. 75, 125–131.

Sithithaworn, P., Ziegler, A.D., Grundy-Warr, C., Andrews, R.H., Petney, T.N., 2012c. Changes to the life cycle of liver flukes: dams, roads, and ponds. Lancet Infect. Dis. 12, 588.

Snidvongs, A., 2006. Vulnerability to climate change related water resource changes and extreme hydrological events in Southeast Asia. In: A Final Report Submitted to Assessments of Impacts and Adaptations to Climate Change (AIACC), Project No. AS 07, Washington D.C.

Sohn, W.M., Yong, T.S., Eom, K.S., Pyo, K.H., Lee, M.Y., Lim, H., Choe, S., Jeong, H.G., Sinuon, M., Socheat, D., Chai, J.Y., 2012. Prevalence of *Opisthorchis viverrini* infection in humans and fish in Kratie Province, Cambodia. Acta Trop. 124, 215–220.

Soukhathammavong, P.A., Sayasone, S., Phongluxa, K., Xayaseng, V., Utzinger, J., Vounatsou, P., Hatz, C., Akkhavong, K., Keiser, J., Odermatt, P., 2012. Low efficacy of single-dose albendazole and mebendazole against hookworm and effect on concomitant helminth infection in Lao PDR. PLoS Neglected Trop. Dis. 6, e1417.

Southgate, D.A., Durnin, J.V., 1970. Calorie conversion factors. An experimental reassessment of the factors used in the calculation of the energy value of human diets. Br. J. Nutr. 24, 517–535.

Sri-Aroon, P., Butraporn, P., Limsomboon, J., Kerdpuech, Y., Kaewpoolsri, M., Kiatsiri, S., 2005. Freshwater mollusks of medical importance in Kalasin Province, northeast Thailand. Southeast Asian J. Trop. Med. Publ. Health 36, 653–657.

Sripa, B., Pairojkul, C., 2008. Cholangiocarcinoma: lessons from Thailand. Curr. Opin. Gastroenterol. 24, 349–356.

Sripa, B., Tangkawattana, S., Laha, T., Kaewkes, S., Mallory, F.F., Smith, J.F., Wilcox, B.A., 2015. Toward integrated opisthorchiasis control in northeast Thailand: the Lawa project. Acta Trop. 141, 361–367.

Srisawangwong, T., Sithithaworn, P., Tesana, S., 1997. Metacercariae isolated from cyprinoid fishes in Khon Kaen District by digestion technic. Southeast Asian J. Trop. Med. Publ. Health 28 (Suppl. 1), 224–226.

Sueblinvong, T., Piyamputra, A., Priyanont, S., Thaithong, S., 1993. Isoenzyme characterization of *Opisthorchis viverrini* from man. Chula. Med. J. 37, 375–385.

Sukontason, K., Piangjai, S., Muangyimpong, Y., Sukontason, K., Methanitikorn, R., Chaithong, U., 1999. Prevalence of trematode metacercariae in cyprinoid fish of Ban Pao district, Chiang Mai Province, northern Thailand. Southeast Asian J. Trop. Med. Publ. Health 30, 365–370.

Suwannahitatorn, P., Klomjit, S., Naaglor, T., Taamasri, P., Rangsin, R., Leelayoova, S., Mungthin, M., 2013. A follow-up study of *Opisthorchis viverrini* infection after the implementation of control program in a rural community, central Thailand. Parasites Vectors 6, 188.

Suwannatrai, A., Pratumchart, K., Suwannatrai, K., Thinkhamrop, K., Chaiyos, J., Kim, C.S., Suwanweerakamtorn, R., Boonmars, T., Wongsaroj, T., Sripa, B., 2017. Modeling impacts of climate change on the potential distribution of the carcinogenic liver fluke, *Opisthorchis viverrini*, in Thailand. Parasitol. Res. 116, 243–250.

Suwannatrai, A., Suwannatrai, K., Haruay, S., Piratae, S., Thammasiri, C., Khampoosa, P., Kulsantiwong, J., Prasopdee, S., Tarbsripair, P., Suwanwerakamtorn, R., Sukchan, S., Boonmars, T., Malone, J.B., Kearney, M.T., Tesana, S., 2011. Effect of soil surface salt on the density and distribution of the snail *Bithynia siamensis goniomphalos* in northeast Thailand. Geospat. Health 5, 183–190.

Suwannatrai, K., Suwannatrai, A., Tabsripair, P., Welbat, J.U., Tangkawattana, S., Cantacessi, C., Mulvenna, J., Tesana, S., Loukas, A., Sotillo, J., 2016. Differential protein expression in the hemolymph of *Bithynia siamensis goniomphalos* infected with *Opisthorchis viverrini*. PLoS Neglected Trop. Dis. 10, e0005104.

Tansurat, P., 1971. Opisthorchiasis. In: Marcial-Rojas, R.A. (Ed.), Pathology of Protozoal and Helminthic Diseases. Williams and Wilkins, Baltimore, pp. 536–545.

Taraschewski, H., Petney, T.N., 2011. Intensivierte Wassernutzung und Globalisierung: Wie Parasiten profitieren. In: Bannert, B., Knopf, K., Vohland, K. (Eds.), Infektionskrankheiten und Biodiversität in anthropogen veränderten Gewässern, Zusammenfassungen des Workshops vom 24 und 25 November 2011 in Berlin. Federal Ministry for Education and Research, Berlin.

Tesana, S., Kaewkes, S., Srisawangwonk, T., Pinlaor, S., 1985. The distribution and density of *Opisthorchis viverrini* metacercariae in cyprinoid fish in Khon Kaen Province. J. Parasitol. Trop. Med. Assoc. Thailand 8, 36–39.

Tesana, S., Khampoosa, P., Piratae, S., Prasopdee, S., Sripanidkulchai, B., 2018. Egg-hatching mechanism of human liver fluke, *Opisthorchis viverrini*: a role for leucine aminopeptidases from the snail host, *Bithynia siamensis goniomphalos*. J. Parasitol. in press.

Thaenkham, U., Blair, D., Nawa, Y., Waikagul, J., 2012. Families Opisthorchiidae and Heterophyidae: are they distinct? Parasitol. Int. 61, 90–93.

Thavorntam, W., Tantemsapya, N., Armstrong, L., 2015. A combination of meteorological and satellite-based drought indices in a better drought assessment and forecasting in Northeast Thailand. Nat. Hazards 77, 1453–1474.

Thompson, J.R., Laizé, C.L.R., Green, A.J., Acreman, M.C., Kingston, D.G., 2014. Climate change uncertainty in environmental flows for the Mekong River. Hydrol. Sci. J. 59, 935–954.

Thu, N.D., Dalsgaard, A., Loan, L.T., Murrell, K.D., 2007. Survey for zoonotic liver and intestinal trematode metacercariae in cultured and wild fish in an Giang Province, Vietnam. Kor. J. Parasitol. 45, 45–54.

Touch, S., Komalamisra, C., Radomyos, P., Waikagul, J., 2009. Discovery of *Opisthorchis viverrini* metacercariae in freshwater fish in southern Cambodia. Acta Trop. 111, 108–113.

Traub, R.J., Macaranas, J., Mungthin, M., Leelayoova, S., Cribb, T., Murrell, K.D., Thompson, R.C., 2009. A new PCR-based approach indicates the range of *Clonorchis sinensis* now extends to Central Thailand. PLoS Neglected Trop. Dis. 3, e367.

Upatham, E.S., Sukhapanth, N., 1980. Field studies on the bionomics of *Bithynia siamensis siamensis* and the transmission of *Opisthorchis viverrini* in Bangna, Bangkok, Thailand. Southeast Asian J. Trop. Med. Publ. Health 11, 355–358.

Upatham, E.S., Viyanant, V., Kurathong, S., Rojborwonwitaya, J., Brockelman, W.Y., Ardsungnoen, S., Lee, P., Vajrasthira, S., 1984. Relationship between prevalence and intensity of *Opisthorchis viverrini* infection, and clinical symptoms and signs in a rural community in north-east Thailand. Bull. World Health Organ. 62, 451–461.

van Vliet, M.T.H., Franssen, W.H.P., Yearsley, J.R., Ludwig, F., Haddeland, I., Lettenmaier, D.P., Kabat, P., 2013. Global river discharge and water temperature under climate change. Global Environ. Change 23, 450–464.
Van Zalinge, N., Degen, P., Pongsri, S., Nuov, S., Jensen, J.G., Nguyen, V.H., Choulamany, X., 2004. The Mekong river system. In: Welcome, R.L., Petr, T. (Eds.), Proceedings of the Second International Symposium on the Management of Large Rivers for Fisheries. RAP Publication 2004, FAO Regional Office for Asia and the Pacific, Bangkok, Thailand, pp. 335–357.
Veciana, M., Chaisiri, K., Morand, S., Ribas, A., 2012. Helminths of the Asian house shrew Suncus murinus from Cambodia. Cambodian J. Nat. Hist 2, 115–122.
Vichasri, S., Viyanant, V., Upatham, E.S., 1982. *Opisthorchis viverrini*: intensity and rates of infection in cyprinoid fish from an endemic focus in Northeast Thailand. Southeast Asian J. Trop. Med. Publ. Health 13, 138–141.
Viranuvatti, V., Stitnimankarn, T., 1972. Liver fluke infection and infestation in Southeast Asia. Prog. Liver Dis. 4, 537–547.
Vityakon, P., 2007. Degradation and restoration of sandy soils under different agricultural land uses in northeast Thailand: a review. Land Degrad. Dev. 18, 567–577.
Wang, Y.C., Feng, C.C., Sithithaworn, P., Feng, Y., Petney, T.N., 2011. How do snails meet fish? Landscape perspective needed to study parasite prevalence. EcoHealth 8, 258–260.
Wang, Y.C., Ho, R.C., Feng, C.C., Namsanor, J., Sithithaworn, P., 2015. An ecological study of *Bithynia* snails, the first intermediate host of *Opisthorchis viverrini* in northeast Thailand. Acta Trop. 141, 244–252.
Wang, Y.C., Yuen, R., Feng, C.C., Sithithaworn, P., Kim, I.H., 2017. Assessing the role of landscape connectivity on *Opisthorchis viverrini* transmission dynamics. Parasitol. Int. 66, 402–412.
Wattanakij, N., Thavorntam, W., Mongkolsawat, C., 2006. Analyzing spatial pattern of drought in the Northeast of Thailand using multi-temporal Standardized Precipitation Index (SPI). In: Proceedings of the 27th ASEAN Conference on Remote Sensing. ACRS 2006, Ulaanbaatar, Mongolia, pp. 1221–1226.
Wenz-Mucke, A., Sithithaworn, P., Petney, T.N., Taraschewski, H., 2013. Human contact influences the foraging behaviour and parasite community in long-tailed macaques. Parasitology 140, 709–718.
Wykoff, D.E., Harinasuta, C., Juttijudata, P., Winn, M.M., 1965. *Opisthorchis viverrini* in Thailand—the life cycle and comparison with *O. felineus*. J. Parasitol. 51, 207–214.
Xu, B., Gong, P., Seto, E., Liang, S., Yang, C., Wen, S., Qiu, D., Gu, X., Spear, R.C., 2006. A spatial-temporal model for assessing the effects of intervillage connectivity in schistosomiasis transmission. Ann. Assoc. Am. Geogr. 96, 31–46.
Yin, M., Liu, X., Xu, B., Huang, J., Zheng, Q., Yang, Z., Feng, Z., Han, Z.G., Hu, W., 2016. Genetic variation between *Schistosoma japonicum* lineages from lake and mountainous regions in China revealed by resequencing whole genomes. Acta Trop. 161, 79–85.
Yong, T.S., Shin, E.H., Chai, J.Y., Sohn, W.M., Eom, K.S., Lee, D.M., Park, K., Jeoung, H.G., Hoang, E.H., Lee, Y.H., Woo, H.J., Lee, J.H., Kang, S.I., Cha, J.K., Lee, K.H., Yoon, C.H., Sinuon, M., Socheat, D., 2012. High prevalence of *Opisthorchis viverrini* infection in a riparian population in Takeo Province, Cambodia. Kor. J. Parasitol. 50, 173–176.
Yoshino, T.P., Lodes, M.J., Rege, A.A., Chappell, C.L., 1993. Proteinase activity in miracidia, transformation excretory-secretory products, and primary sporocysts of *Schistosoma mansoni*. J. Parasitol. 79, 23–31.

Zadesenets, K.S., Katokhin, A.V., Mordvinov, V.A., Rubtsov, N.B., 2012. Comparative cytogenetics of opisthorchid species (Trematoda, Opisthorchiidae). Parasitol. Int. 61, 87–89.

Zahoor, Z., Davies, A.J., Kirk, R.S., Rollinson, D., Walker, A.J., 2010. Larval excretory-secretory products from the parasite *Schistosoma mansoni* modulate HSP70 protein expression in defence cells of its snail host, *Biomphalaria glabrata*. Cell Stress Chaperones 15, 639–650.

Zeigler, R.S., Puckridge, D.W., 1995. Improving sustainable productivity in rice-based rainfed lowland systems of South and Southeast Asia. GeoJournal 35, 307–324.

Zhang, X.L., Ali, M.A., 2008. Ficolins: structure, function and associated diseases. Adv. Exp. Med. Biol. 632, 105–115.

Ziegler, A.D., 2012. Water management: reduce urban flood vulnerability. Nature 481, 145.

Ziegler, A.D., Petney, T.N., Grundy-Warr, C., Andrews, R.H., Baird, I.G., Wasson, R.J., Sithithaworn, P., 2013. Dams and disease triggers on the lower Mekong river. PLoS Neglected Trop. Dis. 7, e2166.

CHAPTER TWO

Epidemiology of *Opisthorchis viverrini* Infection

Apiporn Suwannatrai*[,1], Prasert Saichua[§,¶] and Melissa Haswell[||]
*Department of Parasitology, Faculty of Medicine, Khon Kaen University, Khon Kaen, Thailand
[§]Tropical Medicine Graduate Program, Faculty of Medicine, Khon Kaen University, Khon Kaen, Thailand
[¶]WHO Collaborating Centre for Research and Control of Opisthorchiasis (Southeast Asian Liver Fluke Disease), Department of Pathology, Faculty of Medicine, Khon Kaen University, Khon Kaen, Thailand
[||]Faculty of Health, School of Public Health and Social Work, Queensland University of Technology, Brisbane, QLD, Australia
[1]Corresponding author: E-mail: apiporn@kku.ac.th

Contents

Abstract

Opisthorchiasis in the Lower Mekong Subregion is a parasitic disease caused by the liver fluke *Opisthorchis viverrini*. This parasite has a well-documented distribution in Thailand, Lao PDR, Cambodia, Myanmar and Southern Vietnam. In this chapter, we describe the current knowledge of the epidemiology of *O. viverrini* infection, highlighting advances in control efforts made in the last four decades in Thailand and identifying ongoing gaps in our epidemiological knowledge which need to be filled to support efforts to permanently overcome the heavy morbidity and mortality burden caused by these parasites within their endemic regions.

Advances in Parasitology, Volume 101
ISSN 0065-308X
https://doi.org/10.1016/bs.apar.2018.05.002

1. INTRODUCTION

Opisthorchiasis and clonorchiasis are food-borne trematode infections caused by liver flukes in the family of Opisthorchiidae (*Opisthorchis viverrini*, *Opisthorchis felineus* and *Clonorchis sinensis*). Despite their severe risks to health and the availability of a highly effective drug, pranziquantel, for treatment and substantial efforts in preventative health education since the mid-1980s, these infections persist as a significant neglected tropical disease risk to the 700 million residents of endemic areas within East and Southeast Asia, including Siberia, as well as some European countries (Sithithaworn et al., 2012; Sripa et al., 2010). Each species has a distinct geographic distribution within these regions, with an estimated 10 million human *O. viverrini* infections in the Lower Mekong Basin (Thailand, Lao PDR, Cambodia, Myanmar and Southern Vietnam) (Aung et al., 2017; Sithithaworn et al., 2012) (Fig. 1), eight million human *C. sinensis* infections in China, Northern Vietnam, South Korea and Eastern Russia (Qian et al., 2016) and 1.6 million human *O. felineus* infections in Western Siberia and other parts of the Russian Federation (Fedorova et al., 2018). Increased population migration and tourism may potentially extend the geographical distribution of infections, and possibly local transmission, to other regions and countries (Pumidonming et al., 2018). However, many studies indicate that *O. viverrini* is a species complex and that there is substantial genetic and biological variation between areas. This means that there could be spatial variation in the epidemiology based on these factors (Petney et al., 2016; Saijuntha et al., 2007, 2008; Sithithaworn et al., 2015). Although the frequency of infection in reservoir hosts, such as canines and felines, has been proposed to have slight association with human opisthorchiasis, animal reservoirs may have significant implications for mass drug treatment control opisthorchiasis programs that target only human infections (Chai et al., 2005; Sithithaworn and Haswell-Elkins, 2003). Felines are the most important animal reservoir host of human opisthorchiasis, especially in opisthorchiasis endemic areas (Aunpromma et al., 2012). Feline infection could enable the persistence of the transmission cycle in the environment and may limit the effectiveness of human sanitation in preventing infection reaching cyprinid fish (Jongsuksuntigul and Imsomboon, 2003). More details about animal reservoir hosts of *O. viverrini* are presented in Chapter III of this volume. Importantly, however, active transmission of these parasites is limited to areas where the distributions of snail and fish intermediate hosts, animal

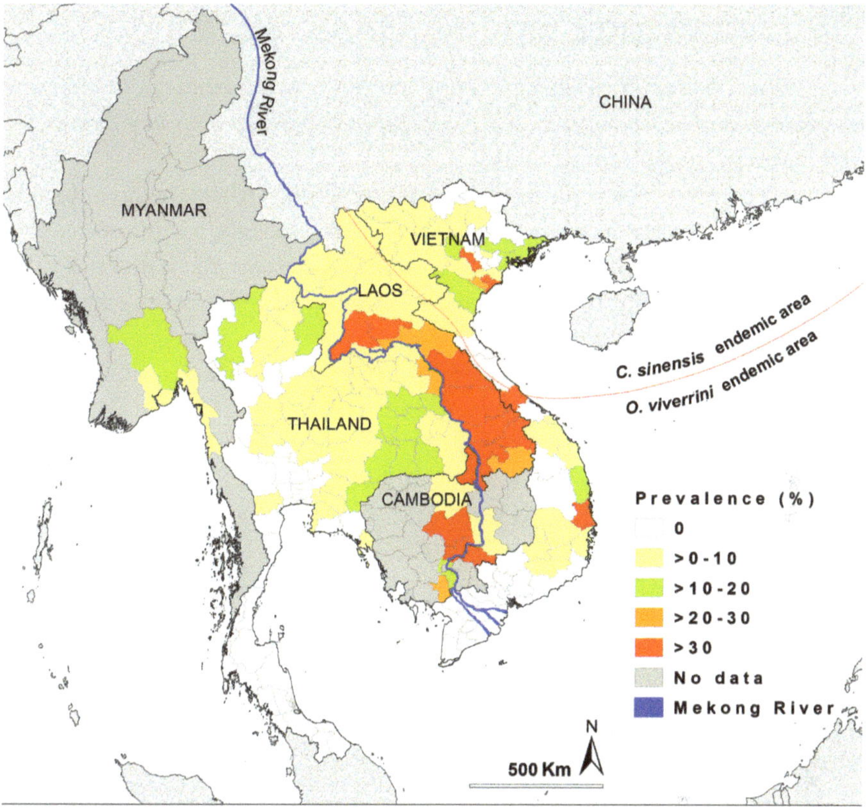

Figure 1 The current prevalence of *Opisthorchis viverrini* and *Clonorchis sinensis* in the Lower Mekong Subregion, including Thailand, Lao PDR, Cambodia, Myanmar and Southern Vietnam. *Modified from Sripa, B., Brindley, P.J., Mulvenna, J., Laha, T., Smout, M.J., Mairiang, E., Bethony, J.M., Loukas, A., 2012. The tumorigenic liver fluke Opisthorchis viverrini–multiple pathways to cancer. Trends. Parasitol. 28, 395-407.*

reservoir, parasites genetic variation, sanitation, human behaviours and other environmental conditions favour transmission to humans (Sithithaworn and Haswell-Elkins, 2003; Upatham, 1976).

This chapter reviews the current knowledge regarding the epidemiology of *O. viverrini* infection, highlighting advances in control efforts made in the last four decades in Thailand and identifying ongoing gaps in our epidemiological knowledge which research can fill to support efforts to permanently overcome the heavy morbidity and mortality burden that this parasite causes in endemic regions.

2. PREVALENCE OF HUMAN INFECTION

Note: When critically reviewing and comparing findings of epidemiological studies of *O. viverrini*, it is important to consider four key factors that may have an impact on accuracy:

1. the potential for self-selection in sampling (those who consider themselves likely to be infected may be more likely to provide a stool sample for analysis than those who do not consider themselves at risk);
2. the age distribution of participants (lower prevalence is expected if children are included);
3. the method of processing and examining the samples as sensitivities at various levels of infection vary (Sithithaworn et al., 1991; Sripa et al., 2011)
4. the relative prevalence of intestinal fluke infections whose eggs are difficult to distinguish from those of other liver flukes, leading to potential overestimation (Kaewkes et al., 1991).

Approximately eight million of the estimated 10 million people infected with *O. viverrini* live in Thailand, and most reside in the Northeast (Isan) region which retains cultural and linguistic similarities with neighbouring Laos (Sithithaworn et al., 2012; Sripa et al., 2010). Sadun (1955) was the first to report high infection rates, up to 100%, in some villages of Northeast Thailand located close to large water bodies (Sadun, 1955). The first national survey conducted in Thailand in 1980–81 reported an overall prevalence of *O. viverrini* of 14.7%, with the highest prevalence found in the Northeast (34.6%), followed by the Central region (6.3%), the North (5.6%) and South (0.01%) (Jongsuksuntigul and Imsomboon, 2003). It should be noted that reported prevalences are likely underestimates due to the limited sensitivity of the Kato-Katz smear method in detecting light infections.

These findings stimulated the implementation of a national opisthorchiasis control program from 1984 to 2001, covering the Northeast, North and some Central provinces. This program focused on mass screening and the treatment of infected individuals with the anthelmintic drug, praziquantel (single dose of 40 mg per kg body weight). Praziquantel became widely available in the early 1980s and has remained the only drug of choice with 91% efficacy against liver fluke infection (Jongsuksuntigul and Imsomboon, 2003; Sripa, 2018). By the year 1991 the national prevalence of *O. viverrini* infection had fallen from 15.2% in the 1991 to 11.8% in 1996 and 9.6% in 2001 (Jongsuksuntigul and Imsomboon, 2003) (Fig. 2).

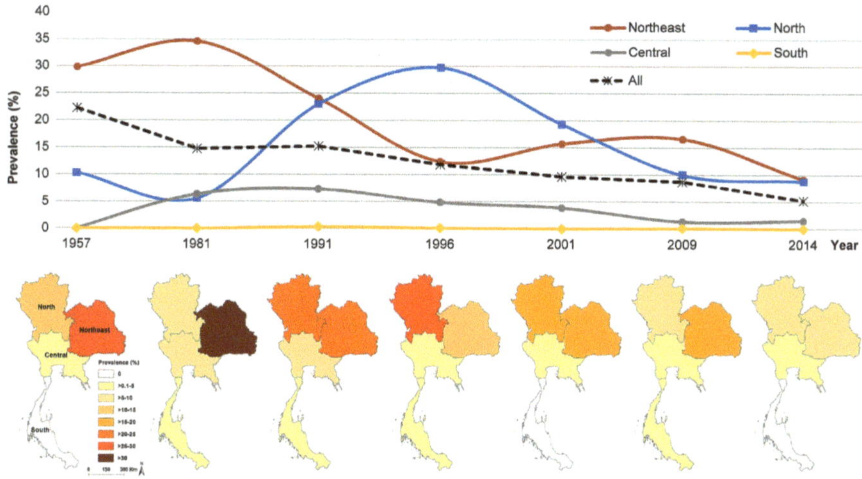

Figure 2 Changes in *Opisthorchis viverrini* prevalence varied greatly by region in Thailand from 1957 to 2014. *Modified from Sithithaworn, P., Andrews, R.H., Nguyen, V.D., Wongsaroj, T., Sinuon, M., Odermatt, P., Nawa, Y., Liang, S., Brindley, P.J., Sripa, B., 2012. The current status of opisthorchiasis and clonorchiasis in the Mekong Basin. Parasitol. Int. 61, 10–16.*

Remarkably prevalence estimates increased in Northern Thailand from 1981 to 1991, exceeding estimates for the Northeast in 1996, but declined to relative parity with the Northeast since 2001 (Fig. 2). While the reasons for this apparent increase remain unclear, a recent study by Pumidonming et al. (2018) identified a significant contribution of migrants from endemic communities in the Northeast to the presence of *O. viverrini* infection in communities in the lower Northern provinces that may have been over-sampled in the national survey (Pumidonming et al., 2018).

Further national reductions in the prevalence of opisthorchiasis accompanied ongoing control efforts in 2009 (8.7%) and 2014 (5.2%) (MOPH, 2014; Wongsaroj et al., 2014). However, many individual provinces within the North and Northeast had retained prevalences above 15% in 2009 (Fig. 2), notably in Nakhon Phanom (60.8%), Sisaket (38.6%) and Amnat Charoen (32.6%). By 2014, the prevalences in these provinces had been reduced substantially, with Nakhon Phanom being the only province with a prevalence above 20% (Fig. 3) (MOPH, 2014; Wongsaroj et al., 2014). While significant achievements have been made nationally and provincially, substantial community-level variation in prevalence persists today.

Opisthorchiasis was first reported in Lao PDR in 1929 when Bedier and Chesnau published the first diagnosed cases of infection with *Opisthorchis* sp.

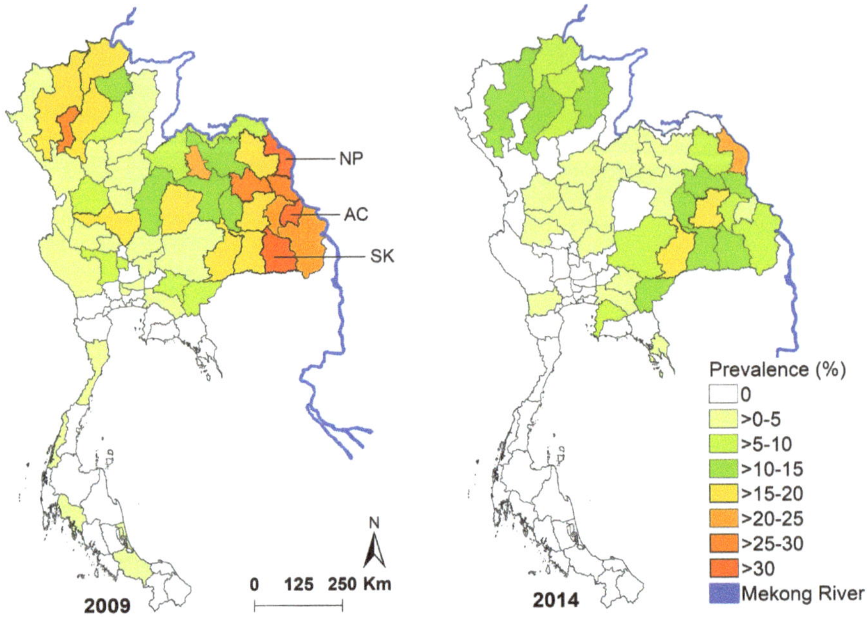

Figure 3 The national survey prevalence of *Opisthorchis viverrini* varied greatly by province in Thailand in 2009 and 2014; *NP*, Nakhon Phanom Province, *AC*, Amnat Charoen Province; *SK*, Sisaket Province.

in Vientiane and Thakhek (Bedier and Chesneau, 1929). Today, an estimated two million people in Lao PDR are infected with *O. viverrini* (Sithithaworn et al., 2012), with most residing in the Central and Southern regions in lowland areas along the Mekong River Basin (Phongluxa et al., 2013; Sayasone et al., 2011).

The earliest epidemiological survey of this parasite in Lao PDR, conducted in 1971 in the Vientiane Province, revealed a prevalence of 46.5% among 2493 residents tested (Sornmani et al., 1974). In 1996 a study in Khammouane Province identified *O. viverrini* infection among 56.7% of 669 residents (Vannachone et al., 1998), similar to a study conducted later which found prevalences of 52.9% in two rural villages and 60.7% in one urban community (Kobayashi et al., 2000).

The first national survey of *O. viverrini* infection in Lao PDR was conducted in 2000–02. This examined faecal samples from 29,846 schoolchildren attending 317 primary schools within 17 provinces and the Vientiane Municipality. The overall prevalence of *O. viverrini* infection among the schoolchildren was 10.9%, with the highest prevalence found in schools in the provinces along the Mekong River, namely Khammouane (32.2%), Savannakhet (25.9%) and Saravane (21.5%) (Rim et al., 2003). Because the prevalence of infection is often lower among children than among adults, there may be underestimates of community-wide prevalences in these areas.

The importance of proximity to the Mekong River was further confirmed by Chai et al. (2007), whose study found *O. viverrini* or minute intestinal fluke eggs (species not distinguished) in the stools of 67.1% of residents living alongside the river. Similarly a survey of 51 villages in Champasack Province found 61.1% of residents with an *O. viverrini* infection (Forrer et al., 2012). Infections may be even more widespread in some communities in the Southern provinces along the river, with prevalences more than 90% observed among residents of Khong and Mounlapamok in Champasack Province (Sayasone et al., 2011). In contrast, a study in the mountainous area of Paksong District reported infection in just 5.7% of residents (Sayasone et al., 2011).

Recent studies have suggested that the prevalence of opisthorchiasis in Lao PDR has not declined even in its capital city where 51%–53% of residents were found to be infected (Nakamura, 2017), and its range may be expanding (Fig. 4). Studies in Don Khon and Don Som islands in Khong District revealed *O. viverrini* infection among 60.7% of 994 participants (Vonghachack et al., 2017).

In summary, these results suggest that opisthorchiasis remains highly endemic in the lowland areas along Mekong River Basin in Lao PDR (Sithithaworn et al., 2012).

Although Cambodia has been identified as an endemic area (Touch et al., 2009), there are few reports on opisthorchiasis in this country. Early evidence was based on the reporting of *C. sinensis* eggs (most likely misidentified *O. viverrini* eggs) (Sithithaworn et al., 2012, Sohn et al., 2011) in stool samples of two recently arrived immigrant Cambodian children in the United States after temporary residence in a refugee camp on the Thai–Cambodian border (Parish, 1985). *Opisthorchis* sp. was first identified in Cambodia in 1994–95 in a schistosomiasis survey in 20 primary schools located along the Mekong River in the Northern province of Kracheh (Stich et al., 1999). *Opisthorchis* infection was also found among schoolchildren in rural and semiurban areas in Central Cambodia, in Stung Treng Province (Sinuon et al., 2003) and in low prevalence (4%) in Kampong Cham Province (Lee et al., 2002). Surveys of schoolchildren and adult residents of riparian villages along the Mekong River in Kratie Province in 2011 also identified a low prevalence of infection (4.6%) (Sohn et al., 2012).

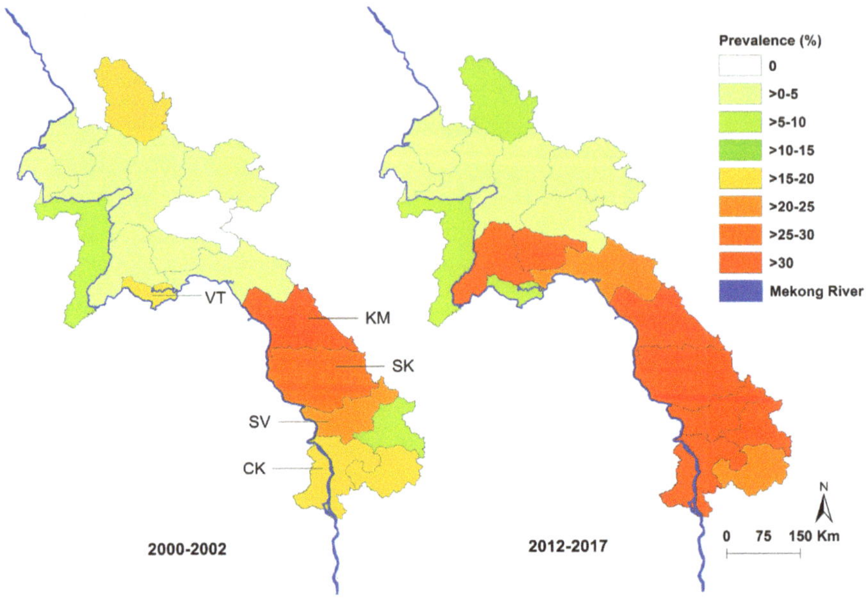

Figure 4 The prevalence of *Opisthorchis viverrini* varied greatly by province in Lao PDR in 2000–2002 and 2012–17; *CK*, Champasack Province; *KM*, Khammouane; *SK*, Savannakhet; *SV*, Saravane; *VT*, Vientiane.

The first large survey in Cambodia was carried out during 2006–12, which examined 16,082 stool samples from residents of 55 villages in five provinces. The overall prevalence of *O. viverrini* infection was 7.7% (1,232/16,082) with province-wide prevalences ranging from 34.8% in Kampong Thom, 33.2% in Kampong Cham, 21.5% in Takeo and 20.2% in Kandal provinces to no infection in Prey Veng Province. District-wide variation was observed, with the highest prevalence (65.1%) found in Kampong Siem District in Kampong Cham Province (Miyamoto et al., 2014) (Fig. 1). Higher prevalences of infection, ranging from 32.4% (Sohn et al., 2011) and 46.4% to 50.6% (Yong et al., 2012) have been found in villages in Prey Kabbas District, Takeo Province.

In summary, these studies suggest that *O. viverrini* infection is present at varying prevalences along the mainstream and tributaries of the Lower Mekong Basin in Cambodia (Yong et al., 2012). No coordinated control efforts have been reported.

Vietnam is the only country known to harbour two human liver fluke species in the family of Opisthorchiidae, with *O. viverrini* highly prevalent in the Central and Southern regions and *C. sinensis* confined to the North

(Sripa et al., 2010; Vo The et al., 2014). A survey conducted in 15 Vietnamese provinces between 1976 and 2000 found that 21% of the population was infected with either *O. viverrini* or *C. sinensis.*

Prevalences ranging from 0.3% to 36.9% have been reported for the Southern provinces in Vietnam, with Phu Yen Province considered to be a 'hotspot' for *O. viverrini* infection (De, 2004; De et al., 2003; Doanh and Nawa, 2016; Sithithaworn et al., 2012). The WHO (2008) reported a prevalence ranging from 0.20% to 40% in six districts in Central Vietnam, with the highest prevalences (10.1%—40.0%) found in Tuy An and Phu My districts. A recent study in lowland areas of Binh Dinh in Central Vietnam reported infection in 11.4% of the 254 residents sampled (Dao et al., 2016) (Fig. 1).

Although *O. viverrini* infection has been identified in Myanmar, very little is known about its distribution and prevalence. One survey reported a prevalence of 9.3% in three regions of Lower Myanmar. The highest prevalence (18.9%) was observed in the Bago Region, followed by Mon State (5%) and the Yangon Region (3.6%) (Aung et al., 2017) (Fig. 1).

3. GEOGRAPHICAL DISTRIBUTION AND RISK MAPPING

As indicated previously, epidemiological studies suggest that *O. viverrini* infection is concentrated most heavily in the Lower Mekong Subregion. Spatial mapping using Getis-Ord statistics confirmed the existence of local clustering of *O. viverrini* infection along the Mekong River (Fig. 5) and 'hotspots' with the highest risk of *O. viverrini* in Northeast Thailand, Central and South Lao PDR, Central Vietnam and North Cambodia. The cold spots, or low risk areas, are confined to North Vietnam that is an endemic area of *C. sinensis.* These 'hotspot' areas receive flows from the Mekong River and its tributaries, creating rich water resources, such as marshes, swamps, lakes and small pools, in the wetland plains where snails and fish thrive (Upatham, 1988).

Recent studies by Suwannatrai et al. (in preparation) used geographical information systems, remote sensing and Maximum Entropy (MaxEnt) algorithm software to generate predictive risk mapping and a distribution pattern for *O. viverrini* infection in response to environmental and climatic factors in Thailand (Fig. 6). This analysis used prevalence data of *O. viverrini* infection from the two nationwide surveys carried out in 2009 and 2014 by the Department of Disease Control, Ministry of Public Health,

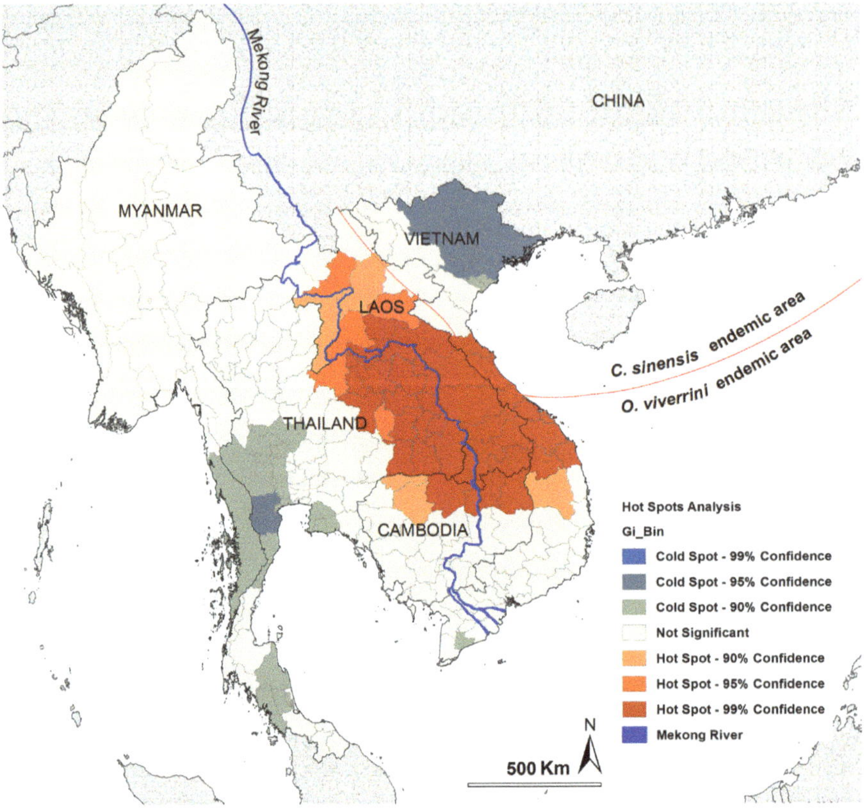

Figure 5 The spatial distribution of *Opisthorchis viverrini* infection prevalence across the Lower Mekong Basin.

Thailand. Environmental data including land surface temperature, normalized difference vegetation index (NDVI) and normalized difference water index (NDWI) were obtained from MODIS satellite data at a resolution of 0.25 km^2 for the time period 1 January 2001 through 31 December 2014. Soil pH and soil texture data were obtained from SoilGrids database (https://soilgrids.org), and land cover data were downloaded from International Steering Committee for Global Mapping (ISCGM) database (http://www.iscgm.org/). Altitude data and 19 bioclimatic variables were obtained from the WorldClim database. All environmental and climatic data were processed using ArcMap 10.5.1. software (Esri Inc., Redlands, CA, USA), whereas MaxEnt version 3.4.1 software (http://biodiversityinformatics.amnh.org/open_source/maxent) was used to construct the model. MaxEnt analysis was run using 25% records for random testing and 75% for training

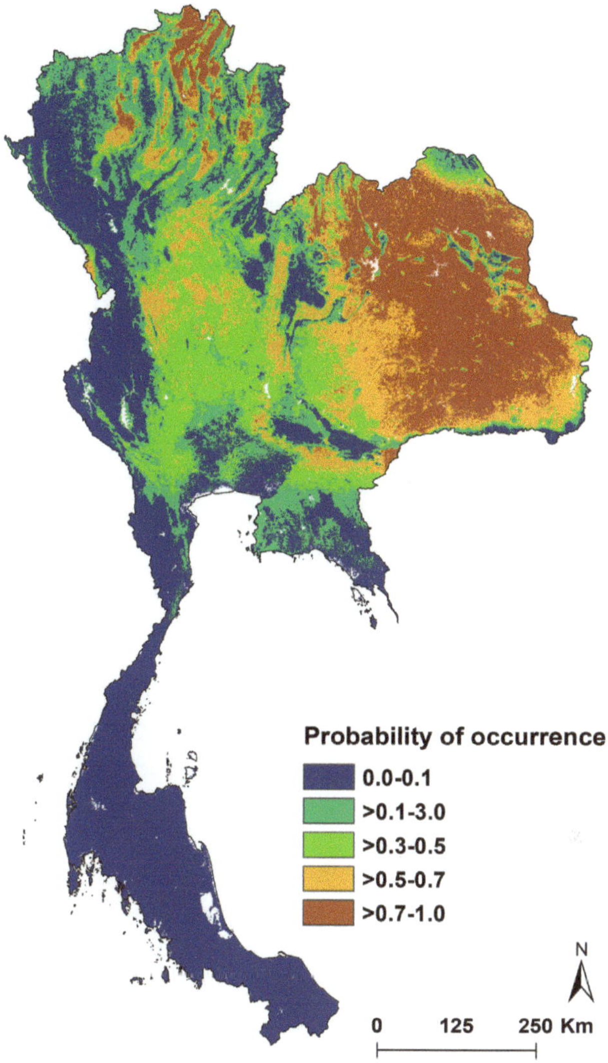

Figure 6 Predicted probability of *Opisthorchis viverrini* occurrence in Thailand.

data to build the model. The area under the curve (AUC) and omission rate were used to validate the models. The training AUC values were greater than 0.8, which indicates an excellent model for *O. viverrini* under environmental and climatic conditions.

This modelling indicated that land cover (paddy field) was the key factor affecting *O. viverrini* distribution in Thailand, followed by temperature seasonality (BIO4) and NDVI. The models predicted that *O. viverrini* is found mainly in Northeast Thailand with spotty distributions in the North (Fig. 6), which is consistent with the trends observed in the many epidemiological studies reviewed here.

4. INCIDENCE AND REINFECTION AFTER TREATMENT

Only a few studies have attempted to measure the incidence of new *O. viverrini* infections. One study over a 2-year period in three villages in Chonnabot District, Khon Kaen, Thailand, found a higher incidence of infection among males than females, particularly in preschoolchildren (aged <5 years) with incidence ranging from 2.1% to 6.2%. The overall incidence in the first year was between 19.4% and 46.0% but decreased to below 20.0% in the second year (Upatham et al., 1985). The three villages in Khon Kaen, Northeast Thailand, reported incidences per 6 months ranging between 1.70% and 25.0% in the year 1993 (Saowakontha et al., 1993). An incidence of 21.6/100 person-years was reported in a rural community of Chachoengsao Province, Central Thailand, in 2002–04 (Rangsin et al., 2009). A follow-up study, after the implementation of a control program in this community in 2007–09, revealed an incidence of *O. viverrini* infection of 21.4/100 person-years (Suwannahitatorn et al., 2013).

Despite extensive studies examining the prevalence of *O. viverrini* infection in many parts of Northeastern Thailand, as well as extensive efforts to eliminate the infection once identified, there are only a few studies reporting reinfection rates after treatment. In 1984, Sornmani et al. compared rates of reinfection in two villages in Nong Wai District after praziquantel treatment (single dose) alone versus praziquantel treatment plus sanitation improvement and health education. The prevalence of *O. viverrini* had returned to pretreatment levels within 1 year (55.1% pretreatment and 54.8% follow-up) in the community that had only received praziquantel treatment. In contrast, the prevalence of infection remained below pretreatment levels (58.3% pretreatment and 36.8% follow-up) within the community that also experienced health education and sanitation improvements, suggesting that multipronged control strategies may stimulate longer lasting reductions (Sornmani et al., 1984a). A similar comparative study was conducted in Nampong Project area, Khon Kaen Province. This revealed a mean monthly reinfection rate of 2.0% and an annual cumulative reinfection rate of 21.5% in two villages where screening and praziquantel treatment as well as health education and sanitation were carried out. In contrast, the monthly reinfection rate was

5.0%, and the annual cumulative rate was 55.5% in a village receiving only praziquantel treatment (Sornmani et al., 1984b). Upatham et al. (1988) observed a return to pretreatment levels within 1 year after treatment (88.4% pretreatment and 87.7% 1 year later) in villages within a heavily infected region (Chonnabot) of Khon Kaen in 1980–82. One recent study in Yasothon Province found a reinfection rate of 10.9% after 12 months among 607 infected villagers in a community in Yasothon Province (Saengsawang et al., 2016). No data regarding reinfection rates of *O. viverrini* after treatment in Cambodia, Lao PDR, Vietnam and Myanmar are available.

5. AGE- AND SEX-RELATED PATTERNS OF INFECTION

Despite the large variation in infection levels, age- and sex-related patterns and transmission dynamics of human opisthorchiasis in Thailand, Lao PDR, Cambodia, Vietnam and Myanmar are similar. Multiple surveys based on the examination of *O. viverrini* eggs in faecal samples have demonstrated that the prevalence and intensity of *O. viverrini* egg output is often lowest among preschoolchildren age groups (below 5 years), increasing among young school-age children (up to 11 years) and slightly rising or plateauing across adolescent, young adult and older age groups (Brockelman et al., 1987; De, 2004; Giboda et al., 1991; Haswell-Elkins et al., 1991; Jongsuksuntigul and Imsomboon, 1997; Kobayashi et al., 2000; Sadun, 1955; Sithithaworn and Haswell-Elkins, 2003) (Fig. 7). These studies found little difference in the prevalence or intensity of infection between males and females.

However, two studies that directly examined worm burden, based on recovery at autopsy (Sithithaworn et al., 1991) and by collecting expelled worms after community-based praziquantel treatment (Haswell-Elkins et al., 1991), observed a much sharper and continuous increase in worm burden throughout all age groups up to 54 years, followed by a decline in elderly age groups. These patterns were not demonstrable by egg counts, possibly because of density-dependent constraints on egg output by adult worms in heavier infections, which is a common phenomenon observed widely among helminth infections, including *O. viverrini* (Elkins et al., 1991; Sithithaworn et al., 1991). Intensity of infection is an important determinant of both transmission and likelihood of clinical consequences (Elkins et al., 1991; Sithithaworn et al., 1991).

The studies discussed previously were undertaken before widespread control efforts in Northeast Thailand (years 1951–2007) (Fig. 7). More

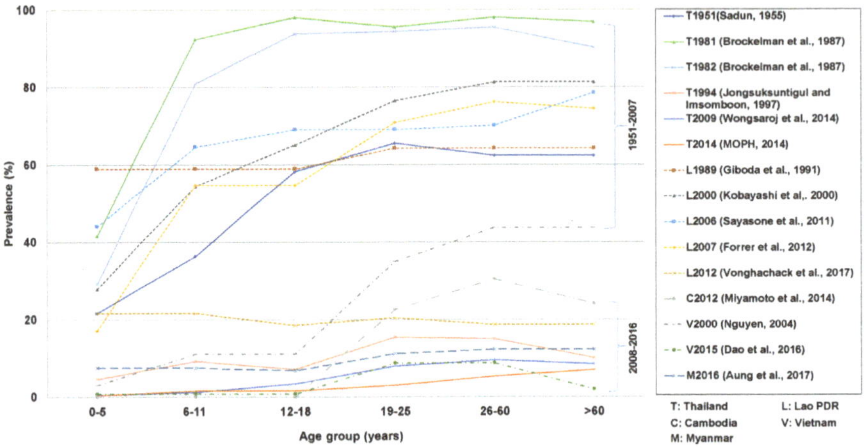

Figure 7 The age pattern of *Opisthorchis viverrini* infection in Thailand, Lao PDR, Cambodia, Myanmar and Southern Vietnam observed during years before (1951–2007) and after (2008–16) significant control efforts. *Modified from Khuntikeo, N., Sithithaworn, P., Loilom, W., Namwat, N., Yongvanit, P., Thinkhamrop, B., Kiatsopit, N., Andrews, R.H., Petney, T.N., 2016. Changing patterns of prevalence in* Opisthorchis viverrini *sensu lato infection in children and adolescents in northeast Thailand. Acta Trop. 164, 469–472.*

recent broad surveys (2008–2016) have reported lower prevalences of *O. viverrini* infection across all age groups (Aung et al., 2017; Dao et al., 2016; Forrer et al., 2012; Khuntikeo et al., 2016; Miyamoto et al., 2014; MOPH, 2014; Sayasone et al., 2011; Vonghachack et al., 2017; Wongsaroj et al., 2014) (Fig. 7). Recent studies also show little to no difference in prevalence and intensity of infection between males and females (Aung et al., 2017; Dao et al., 2016; Kobayashi et al., 2000; Sayasone et al., 2007; Upatham et al., 1984; Vonghachack et al., 2017).

While data from older studies are clearly no longer relevant to areas where extensive control programs have been conducted, these patterns of infection observed before intensive control efforts may be more typical of other areas that have not had significant control efforts applied.

6. PERSISTENT RISK FACTORS OF INFECTION DESPITE CONTROL EFFORTS

As detailed previously, despite significant control efforts, *O. viverrini* remains endemic, with an estimated 10 million residents infected, in rural and suburban communities of the Lower Mekong Subregion, including

Thailand, Lao PDR, Cambodia, Myanmar and Vietnam (Sithithaworn et al., 2012). The mode of and major risk factor for human infection is the consumption of raw or undercooked cyprinid fish containing viable, infective metacercariae. Estimates of increased risk of *O. viverrini* infection among people who report consuming raw cyprinid fish compared with those who do not ranged from 2 to 7 times higher in Thailand and Lao PDR (Rangsin et al., 2009; Suwannahitatorn et al., 2013) to 107 times higher in Central Vietnam (Dao et al., 2016). The substantially lower estimates reported in Thailand and Lao PDR than in Vietnam may result from artifacts of sampling or other differences in the study designs. Alternatively these may result from consumption of Thai and Lao dishes that are not widely recognized as containing raw fish, leading to underreporting and/or the influence of praziquantel treatment and/or improved sanitation in reducing transmission in the two countries leading to reduced likelihood of infection despite consuming raw fish.

There are many local dishes that contain raw fish, and their consumption is a long-standing practice among people who live near wetlands and water bodies, engaging in rice farming, fishing and animal breeding (Dao et al., 2016; Forrer et al., 2012; Grundy-Warr et al., 2012; Miyamoto et al., 2014; Thaewnongiew et al., 2014; Xayaseng et al., 2013). In Northern Thailand, the risk of infection was found to be higher among migrants from the Northeast who have either not had access to treatment to eliminate long-standing infections and/or who experience repeated infections because of local transmission (Pumidonming et al., 2018).

There is a complexity of other risk factors that promote the persistence of *O. viverrini* infection in endemic regions, some of which are related to eating preferences, whereas others (as discussed in Section 3) are environmentally influenced (Sripa et al., 2017). Table 1 summarizes the evidence from multiple studies reporting a higher risk of infection linked to older age, history of diagnosed infection leading to praziquantel treatment and education levels (Table 1). Studies in Northeastern Thai provinces have found males and those aged between 30 and 60 years as major risk groups (Chaiputcha et al., 2015; Chudthaisong et al., 2015; Prakobwong et al., 2017; Thaewnongiew et al., 2014). A similar risk pattern for *O. viverrini* has been observed in Lao PDR (Forrer et al., 2012; Phongluxa et al., 2013; Saiyachak et al., 2016; Yoon et al., 2014) and Central Vietnam (Dao et al., 2016). Miyamoto et al. (2014) reported that adults of working age are most likely to be infected with *O. viverrini* infection in Cambodia. Although a lower prevalence of infection among young children has almost

Table 1 The Risk Factor of *Opisthorchis viverrini* Infection From Selected Publication Between 2009 and 2018

Risk Factor	Finding	Country	References
Age	>50 years	Thailand	Thaewnongiew et al. (2014)
	>35 years	Thailand	Chudthaisong et al. (2015)
	>30 years, increasing risk trend in older age	Thailand	Chaiputcha et al. (2015)
	>40 years	Thailand	Prakobwong et al. (2017)
	>7 years, increasing risk trend in older age	Lao PDR	Forrer et al. (2012)
	>20 years, increasing risk trend in older age	Cambodia	Yong et al. (2012)
Sex	Male	Thailand	Thaewnongiew et al. (2014), Chaiputcha et al. (2015)
	Male	Lao PDR	Yoon et al. (2014), Saiyachak et al. (2016)
	Male	Vietnam	Dao et al. (2016)
Education	Secondary school and below	Thailand	Prakobwong et al. (2017)
	Secondary school and above	Lao PDR	Forrer et al. (2012)
Behaviour	History of *O. viverrini* infection	Thailand	Thaewnongiew et al. (2014), Chudthaisong et al. (2015)
	Raw fish consumption	Thailand	Rangsin et al. (2009), Suwannahitatorn et al. (2013), Thaewnongiew et al. (2014), Chudthaisong et al. (2015), Prakobwong et al. (2017)
	Raw fish consumption	Lao PDR	Tomokawa et al. (2012), Yoon et al. (2014), Saiyachak et al. (2016)
	Raw fish consumption	Vietnam	Dao et al., 2016
	Past use of praziquantel	Thailand	Saengsawang et al., 2013, Thaewnongiew et al., 2014, Chudthaisong et al., 2015
	Proximity to water bodies	Thailand	Thaewnongiew et al. (2014)
	Water resource habitat	Lao PDR	Forrer et al. (2012)
Occupation	Rice farmer	Lao PDR	Forrer et al. (2012)
	Raise animals/fisherman	Lao PDR	Forrer et al. (2012)
Knowledge	Lack of knowledge about praziquantel treatment	Thailand	Saengsawang et al. (2013), Prakobwong et al. (2017)

universally been observed, these recent findings of infection risk increasingly recognized at older ages raise the possibility that schoolchildren and young people may be heeding health education efforts and remain uninfected after treatment more consistently than older people (Khuntikeo et al., 2016).

The contrasting difference in risk associated with education levels at or above secondary level was observed between studies in Thailand (apparently reducing risk) (Chaiputcha et al., 2015; Prakobwong et al., 2017; Yoon et al., 2014) and Lao PDR (apparently increasing risk) (Forrer et al., 2012). Other studies suggest no link between educational level and raw fish consumption or infection (Promthet et al., 2015; Saiyachak et al., 2016). A case–control study in schoolchildren in Lao PDR found that the risk of *O. viverrini* infection depends on the child's experience of eating raw fish, the frequency of eating raw fish by the guardian and maternal education, but not the guardian's knowledge, attitudes or perceptions of liver fluke infection (Tomokawa et al., 2012). These ambiguous results may reflect confounding of ethnicity and cultural practices (Laos and Isan Thai), and varying levels of social and education status in the two countries rather than the education itself.

7. NEW UNDERSTANDING REGARDING THE CONSUMPTION OF RAW FISH

Recent qualitative studies using interviews and focus groups are beginning to provide a new and deeper understanding of the meaning and persistence of raw fish consumption in Thailand and Lao PDR. This information is being integrated in current prevention programs. These studies suggest that a lack of awareness or acceptance that *O. viverrini* infection results from eating raw or undercooked cyprinoid fish continues to persist. There is also evidence that despite this knowledge, the social and cultural importance of these dishes to the Lao and Isan populations may currently take precedence over decisions on whether to consume them.

Long-standing beliefs regarding the ability of added ingredients, such as lime juice, chili or monosodium glutamate, to destroy the infective stage of liver fluke persist. Others appear to have accepted that eating raw fish carries a risk but are able to limit their risk by consuming raw fish only at special events and cultural ceremonies (Kim et al., 2017; Xayaseng et al., 2013). In Vietnam a survey identified that more than 25% of *O. viverrini*-positive participants understood the harmful effects of raw fish consumption but continued to believe that eating raw fish can reduce illness and promote good health (Dao et al., 2016). These misbeliefs are quite stubborn and have contributed to the continuing high prevalence of *O. viverrini* infection in the Lower Mekong Subregion.

Recent studies have shed greater light on the social and economic importance of cyprinid fish sharing in the Lower Mekong Subregion. Local fishermen and farmers catch these fish from water reservoirs, consuming and sharing the fish with their extended family and neighbours (Grundy-Warr et al., 2012; Saenna et al., 2017). This food-sharing activity is symbolic of sharing their cultural identity (Klausner, 1993). Raw fish dishes are particularly valued as part of cultural celebrations, such as wedding ceremonies and during harvesting after a long arduous growing season (Suwannahitatorn et al., 2013).

Social network analysis from Lawa Lake in Khon Kaen Province in Thailand revealed food sharing between household in villages with both high and low prevalence of *O. viverrini* infection (Phimpraphai et al., 2017). Another study of raw fish sharing from two villages located along the Chi River showed that the degree of food sharing was linked to an estimated 7% increased risk of infection (Saenna et al., 2017). Improved transportation in the Lower Mekong Subregion in recent years has led to the marketing of cyprinid fish by fishermen to middlemen who then distribute them over a much wider area than was previously possible (Kim et al., 2017). The economic potential of selling and distributing cyprinid fish from areas with *O. viverrini* transmission to other areas, coupled with a food-sharing culture, may lead to spreading infection beyond previous boundaries.

Another complicating factor that may be driving the persistence of *O. viverrini* infection is the anthelmintic drug, praziquantel. There is no doubt that the widespread introduction of praziquantel treatment within national liver fluke control programs together with health education in Thailand from 1980 to 2000 (Sripa et al., 2015) played a critical role in the successful reduction in prevalence from 34.6% to 15.7%, but not elimination, in the Northeast of the country (Jongsuksuntigul and Imsomboon, 2003). Reinfection after liver fluke treatment related to persistent raw fish eating has allowed transmission to continue, albeit at lower levels (Saengsawang et al., 2016). Ironically some research suggests that knowledge and availability of praziquantel may be a deterrent to permanent avoidance of raw fish consumption as some individuals adopt a 'personal' strategy of eating raw fish then seeking praziquantel treatment (Grundy-Warr et al., 2012;

Saengsawang et al., 2016). Furthermore a study in Yasothon Province, Northeast Thailand, found misconceptions about a protective effect of praziquantel against reinfection as a risk factor linked to infection (Saengsawang et al., 2013).

In summary, this section has reviewed the evidence that the major risk factor for *O. viverrini* infection is raw fish consumption, which is influenced by a strong link with important cultural and social practices and preferences that may be most persistent among males in older age groups. Furthermore misconceptions, incomplete knowledge and attitudes regarding the anthelmintic drug praziquantel may be deterring permanent behavioural change and disease eradication. These findings provide critical information that needs to be addressed in efforts towards long-term, sustainable liver fluke control.

8. INTERMEDIATE HOSTS AND ENVIRONMENTAL DETERMINANTS

As mentioned previously, environmental factors influencing the life cycle stages of *O. viverrini* also influence the persistence of the parasite in humans. *Bithynia* sp. snails serve as the first intermediate host of *O. viverrini*, enabling transmission to cyprinid fish which harbour the infective stage to humans (metacercaria). Factors promoting the density of these snails, the likelihood of infected human and reservoir species faeces reaching the water body and the likely survival and success of the free-swimming cercariae that exit the snail, attach and penetrate into the tissue of cyprinid fish to transform into metacercariae determine the potential infective dose to humans (Upatham, 1988). More details about intermediate hosts of *O. viverrini* are presented in Chapter I of this volume.

Humans acquire *O. viverrini* infection by eating raw or undercooked freshwater cyprinid fish containing metacercariae (Upatham and Viyanant, 2003). Sithithaworn et al. (1997) reported strong seasonal variation in the intensity of *O. viverrini* metacercariae in cyprinid fish in Northeast Thailand. The highest intensities were found during the later part of the rainy season leading to winter, from August to January, and the lowest occurred in the summer and rainy season, from February to July (Sithithaworn et al., 1997). A study in Lao PDR, however, found a high intensity of *O. viverrini* metacercariae in cyprinid fish in the summer months of November to April (Manivong et al., 2009).

9. CLIMATE CHANGE AND INFECTION PREDICTION

Climate change directly affects parasite distribution and transmission via the environmental conditions present, such as increasing temperatures and changing rainfall patterns, and may impact the reproduction and survival of intermediate hosts that are responsive to water temperature (Guernier et al., 2004; McCreesh et al., 2014; Wu et al., 2016). Global temperatures have already risen approximately 1°C and are expected to increase by 0.2°C for each future decade (IPCC, 2007). Thailand is considered highly vulnerable to the effects of climate change, particularly through increased temperatures and unprecedented droughts, such as those observed in the Lower Mekong Basin in 2010 (Marks, 2011). These environmental changes may significantly influence the geographic distribution of opisthorchiasis, particularly because of impacts on its first intermediate host, *Bithynia* snails (Wykoff et al., 1965). Temperature affects mortality, reproduction and the likelihood of *Bithynia* sp. infection by *O. viverrini* (Brockelman et al., 1986). Suwannatrai et al. (2017) recently modelled the impact of climate change on the potential distribution of *O. viverrini* in Thailand. The Hadley Global Environment Model 2—Earth System (HadGEM2-ES) climate change model and the IPCC scenarios A2a for 2050 and 2070 were used as climatic scenarios. The occurrence data on *O. viverrini* in 2009 and 2014 were obtained from the Department of Disease Control, Ministry of Public Health. MaxEnt algorithm software was used to construct an ecological niche model of *O. viverrini* distribution. This modelling suggested that precipitation and minimum temperature are already significantly affecting the distribution of *O. viverrini* within the Northeast and Northern regions. By 2050 the distribution of *O. viverrini* will be particularly affected by precipitation, maximum temperature and mean temperature in the wettest quarter (BIO8), whereas by 2070, precipitation in the coldest quarter (BIO19) and maximum and minimum temperatures are likely to be the most significant factors. It is expected that these climate alterations resulting from global climate change will significantly reduce the geographical distribution of *O. viverrini* in the Northeast region of Thailand (Fig. 8) (Suwannatrai et al., 2017).

10. CONCLUSIONS AND FUTURE PERSPECTIVES

The distribution and transmission of *O. viverrini* remains a major public health problem in the Lower Mekong Basin including Thailand, Lao

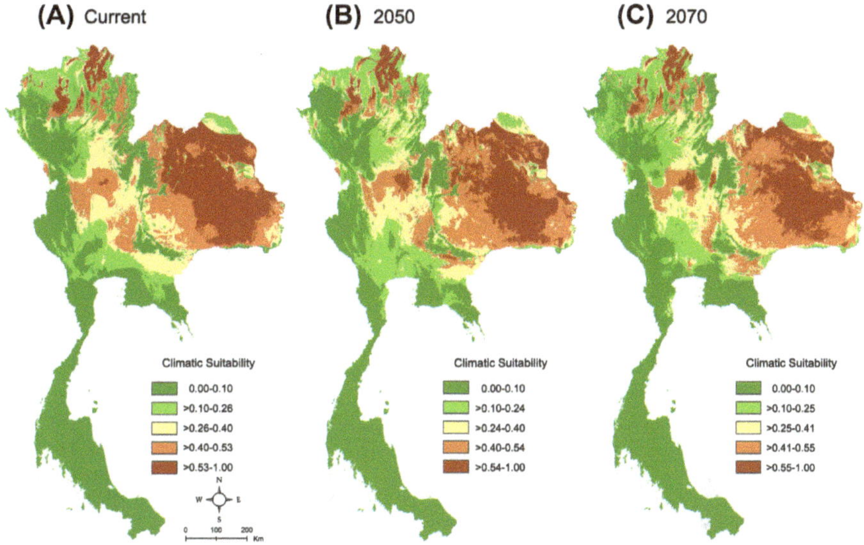

Figure 8 Predicted probability of the *Opisthorchis viverrini* distribution in current (A) and future (2050 (B) and 2070 (C)) scenarios in Thailand based on climatic factors. The future distribution map presented a drastic decrease in suitability distribution of *Opisthorchis viverrini* in the Northeast Thailand. *Modified from Suwannatrai, A., Pratumchart, K., Suwannatrai, K., Thinkhamrop, K., Chaiyos, J., Kim, C.S., Suwanweer-akamtorn, R., Boonmars, T., Wongsaroj, T., Sripa, B., 2017. Modelling impacts of climate change on the potential distribution of the carcinogenic liver fluke,* Opisthorchis viverrini, *in Thailand. Parasitol. Res. 116, 243–250.*

PDR, Cambodia, Myanmar and Southern Vietnam (Sithithaworn et al., 2012; Sripa et al., 2010). There has been a great deal of effort made to control *O. viverrini*, especially in Thailand, leading to a continuous decline in opisthorchiasis over the past four decades. However, the prevalence of *O. viverrini* infection remains in the Northeast and North of the country, particularly in rural areas. The distribution of *O. viverrini* has many related factors, such as environmental and climatic factors, its intermediate hosts, sanitation and eating habits. Educating and exchanging proactive learning in teaching children who are at an appropriate age are likely to play an important role in controlling transmission. This should help to adjust the consumption behaviour and personal hygiene in rural communities. In the past, the *O. viverrini* egg identification problems by fecal microscopy were a limitation in the survey data in the Mekong River Basin where minute intestinal flukes (heterophyid and lecithodendriid) are cotransmitted. It is possible that in the past, *O. viverrini* identification may have included those minute intestinal flukes, which are morphologically similar to eggs, and lead to overestimation of infection rate. The use of standard screening methods with higher sensitivity and specificity to determine the prevalence of *O. viverrini* in human hosts, intermediate hosts and reservoir hosts is urgently required for revealing the true prevalence of this liver fluke. This will help for planning and controlling the disease in highly endemic areas.

REFERENCES

Aung, W.P.P., Htoon, T.T., Tin, H.H., Thinn, K.K., Sanpool, O., Jongthawin, J., Sadaow, L., Phosuk, I., Rodpai, R., Intapan, P.M., Maleewong, W., 2017. First report and molecular identification of *Opisthorchis viverrini* infection in human communities from Lower Myanmar. PLoS One 12, e0177130.

Aunpromma, S., Tangkawattana, P., Papirom, P., Kanjampa, P., Tesana, S., Sripa, B., Tangkawattana, S., 2012. High prevalence of *Opisthorchis viverrini* infection in reservoir hosts in four districts of Khon Kaen Province, an opisthorchiasis endemic area of Thailand. Parasitol. Int. 61, 60–64.

Bedier, E., Chesneau, P., 1929. Distomatose hepatique a *Opisthorchis* in Laos (Vientiane and Thakek). Bull Soc Path Exot. 22.

Brockelman, W.Y., Upatham, E.S., Viyanant, V., Ardsungnoen, S., Chantanawat, R., 1986. Field studies on the transmission of the human liver fluke, *Opisthorchis viverrini*, in northeast Thailand: population changes of the snail intermediate host. Int. J. Parasitol. 16, 545–552.

Brockelman, W.Y., Upatham, E.S., Viyanant, V., Hirunraks, A., 1987. Measurement of incidence of the human liver fluke, *Opisthorchis viverrini*, in northeast Thailand. Trans. R. Soc. Trop. Med. Hyg. 81, 327–335.

Chai, J.Y., Darwin Murrell, K., Lymbery, A.J., 2005. Fish-borne parasitic zoonoses: status and issues. Int. J. Parasitol. 35, 1233–1254.

Chai, J.Y., Han, E.T., Guk, S.M., Shin, E.H., Sohn, W.M., Yong, T.S., Eom, K.S., Lee, K.H., Jeong, H.G., Ryang, Y.S., Hoang, E.H., Phommasack, B., Insisiengmay, B., Lee, S.H., Rim, H.J., 2007. High prevalence of liver and intestinal fluke infections among residents of Savannakhet Province in Laos. Korean J. Parasitol. 45, 213–218.

Chaiputcha, K., Promthet, S., Bradshaw, P., 2015. Prevalence and risk factors for infection by *Opisthorchis viverrini* in an urban area of Mahasarakham province, Northeast Thailand. Asian Pac. J. Cancer Prev 16, 4173–4176.

Chudthaisong, N., Promthet, S., Bradshaw, P., 2015. Risk factors for *Opisthorchis viverrini* Infection in Nong Khai province, Thailand. Asian Pac. J. Cancer Prev. 16, 4593–4596.

Dao, T.T., Bui, T.V., Abatih, E.N., Gabriel, S., Nguyen, T.T., Huynh, Q.H., Nguyen, C.V., Dorny, P., 2016. *Opisthorchis viverrini* infections and associated risk factors in a lowland area of Binh Dinh Province, Central Vietnam. Acta Tropica 157, 151–157.

De, N.V., 2004. Fish-borne trematodes in Vietnam. Southeast Asian J. Trop. Med. Publ. Health 35, 299–301.

De, N.V., Murrell, K.D., Cong le, D., Cam, P.D., Chau le, V., Toan, N.D., Dalsgaard, A., 2003. The food-borne trematode zoonoses of Vietnam. Southeast Asian J. Trop. Med. Publ. Health 34 (Suppl. 1), 12–34.

Doanh, P.N., Nawa, Y., 2016. *Clonorchis sinensis* and *Opisthorchis* spp. in Vietnam: current status and prospects. Trans. R. Soc. Trop. Med. Hyg. 110, 13–20.

Elkins, D.B., Sithithaworn, P., Haswell-Elkins, M., Kaewkes, S., Awacharagan, P., Wongratanacheewin, S., 1991. *Opisthorchis viverrini*: relationships between egg counts, worms recovered and antibody levels within an endemic community in northeast Thailand. Parasitology 102 (Pt. 2), 283–288.

Fedorova, O.S., Fedotova, M.M., Sokolova, T.S., Golovach, E.A., Kovshirina, Y.V., Ageeva, T.S., Kovshirina, A.E., Kobyakova, O.S., Ogorodova, L.M., Odermatt, P., 2018. *Opisthorchis felineus* infection prevalence in Western Siberia: a review of Russian literature. Acta Trop. 178, 196–204.

Forrer, A., Sayasone, S., Vounatsou, P., Vonghachack, Y., Bouakhasith, D., Vogt, S., Glaser, R., Utzinger, J., Akkhavong, K., Odermatt, P., 2012. Spatial distribution of, and risk factors for, *Opisthorchis viverrini* infection in southern Lao PDR. PLoS Neglected Trop. Dis. 6, e1481.

Giboda, M., Ditrich, O., Scholz, T., Viengsay, T., Bouaphanh, S., 1991. Current status of food-borne parasitic zoonoses in Laos. Southeast Asian J. Trop. Med. Publ. Health 22 (Suppl.), 56–61.

Grundy-Warr, C., Andrews, R.H., Sithithaworn, P., Petney, T.N., Sripa, B., Laithavewat, L., Ziegler, A.D., 2012. Raw attitudes, wetland cultures, life-cycles: socio-cultural dynamics relating to *Opisthorchis viverrini* in the Mekong Basin. Parasitol. Int. 61, 65–70.

Guernier, V., Hochberg, M.E., Guegan, J.F., 2004. Ecology drives the worldwide distribution of human diseases. PLoS Biol. 2, e141.

Haswell-Elkins, M.R., Elkins, D.B., Sithithaworn, P., Treesarawat, P., Kaewkes, S., 1991. Distribution patterns of *Opisthorchis viverrini* within a human community. Parasitology 103 (Pt. 1), 97–101.

IPCC, 2007. Climatic change 2007: the physical science basis. In: Solomon, S., Qin, D., Manning, M., Chen, Z., Marquis, M., Averyt, K.B., Tignor, M., Miller, H.L. (Eds.), Contribution of Working Group I to Fourth Assessment Report of the Intergovernmental Panel on Climate Change (IPCC) (Cambridge, United Kingdom and New York, NY, USA).

Jongsuksuntigul, P., Imsomboon, T., 1997. The impact of a decade long opisthorchiasis control program in northeastern Thailand. Southeast Asian J. Trop. Med. Publ. Health 28, 551–557.

Jongsuksuntigul, P., Imsomboon, T., 2003. Opisthorchiasis control in Thailand. Acta Trop. 88, 229–232.

Kaewkes, S., Elkins, D.B., Sithithaworn, P., Haswell-Elkins, M.R., 1991. Comparative studies on the morphology of the eggs of *Opisthorchis viverrini* and lecithodendriid trematodes. Southeast Asian J. Trop. Med. Publ. Health 22, 623–630.

Khuntikeo, N., Sithithaworn, P., Loilom, W., Namwat, N., Yongvanit, P., Thinkhamrop, B., Kiatsopit, N., Andrews, R.H., Petney, T.N., 2016. Changing patterns of prevalence in *Opisthorchis viverrini* sensu lato infection in children and adolescents in northeast Thailand. Acta Trop. 164, 469–472.

Kim, C.S., Smith, J.F., Suwannatrai, A., Echaubard, P., Wilcox, B., Kaewkes, S., Sithithaworn, P., Sripa, B., 2017. Role of socio-cultural and economic factors in cyprinid fish distribution networks and consumption in Lawa Lake region, Northeast Thailand: novel perspectives on *Opisthorchis viverrini* transmission dynamics. Acta Tropica 170, 85–94.

Klausner, W.J., 1993. Reflections on Thai Culture: Collected Writings. The Siam Society, Bangkok.

Kobayashi, J., Vannachone, B., Sato, Y., Manivong, K., Nambanya, S., Inthakone, S., 2000. An epidemiological study on *Opisthorchis viverrini* infection in Lao villages. Southeast Asian J. Trop. Med. Publ. Health 31, 128–132.

Lee, K.J., Bae, Y.T., Kim, D.H., Deung, Y.K., Ryang, Y.S., Kim, H.J., Im, K.I., Yong, T.S., 2002. Status of intestinal parasites infection among primary school children in Kampongcham, Cambodia. Kor. J. Parasitol. 40, 153–155.

Manivong, L., Komalamisra, C., Waikagul, J., Radomyos, P., 2009. *Opisthorchis viverrini* metacercariae in cyprinoid fish from three rivers in Khammouane Province, Lao PDR. J. Trop. Med. Parasitol. 32, 23–29.

Marks, D., 2011. Climate change and Thailand: impact and response. Contemp Southeast Asia 33 (2), 229–258.
McCreesh, N., Arinaitwe, M., Arineitwe, W., Tukahebwa, E.M., Booth, M., 2014. Effect of water temperature and population density on the population dynamics of *Schistosoma mansoni* intermediate host snails. Parasit Vectors 7, 503.
Miyamoto, K., Kirinoki, M., Matsuda, H., Hayashi, N., Chigusa, Y., Sinuon, M., Chuor, C.M., Kitikoon, V., 2014. Field survey focused on *Opisthorchis viverrini* infection in five provinces of Cambodia. Parasitol. Int. 63, 366–373.
MOPH, 2014. National Survey on Helminthiasis in Thailand. Department of Disease Control, Ministry of Public Health, Bangkok, Thailand.
Nakamura, S., 2017. Present situation of opisthorchiasis in Vientiane capital, Lao peoples' democratic republic. Nihon Eiseigaku Zasshi 72, 101–105.
Parish, R.A., 1985. Intestinal parasites in Southeast Asian refugee children. West. J. Med. 143, 47–49.
Petney, T.N., Sithithaworn, P., Andrews, R.H., Webster, J.P., 2016. Foodborne trematodes: a diverse and challenging group of neglected parasites. Trans. R. Soc. Trop. Med. Hyg. 110, 1–3.
Phimpraphai, W., Tangkawattana, S., Sereerak, P., Kasemsuwan, S., Sripa, B., 2017. Social network analysis of food sharing among households in opisthorchiasis endemic villages of Lawa Lake, Thailand. Acta Tropica 169, 150–156.
Phongluxa, K., Xayaseng, V., Vonghachack, Y., Akkhavong, K., van Eeuwijk, P., Odermatt, P., 2013. Helminth infection in southern Laos: high prevalence and low awareness. Parasit Vectors 6, 328.
Prakobwong, S., Suwannatrai, A., Sancomerang, A., Chaipibool, S., Siriwechtumrong, N., 2017. A large scale study of the epidemiology and risk factors for the carcinogenic liver fuke *Opisthorchis viverrini* in Udon Thani province, Thailand. Asian Pac. J. Cancer Prev. 18, 2853–2860.
Promthet, P., Kessomboon, P., Promthet, S., 2015. Community-Based health education and communication model development for opisthorchiasis prevention in a high risk area, Khon Kaen Province, Thailand. Asian Pac. J. Cancer Prev. 16, 7789–7794.
Pumidonming, W., Katahira, H., Igarashi, M., Salman, D., Abdelbaset, A.E., Sangkaeo, K., 2018. Potential risk of a liver fluke *Opisthorchis viverrini* infection brought by immigrants from prevalent areas: a case study in the lower Northern Thailand. Acta Trop. 178, 213–218.
Qian, M.B., Utzinger, J., Keiser, J., Zhou, X.N., 2016. Clonorchiasis. Lancet 387, 800–810.
Rangsin, R., Mungthin, M., Taamasri, P., Mongklon, S., Aimpun, P., Naaglor, T., Leelayoova, S., 2009. Incidence and risk factors of *Opisthorchis viverrini* infections in a rural community in Thailand. Am. J. Trop. Med. Hyg. 81, 152–155.
Rim, H.J., Chai, J.Y., Min, D.Y., Cho, S.Y., Eom, K.S., Hong, S.J., Sohn, W.M., Yong, T.S., Deodato, G., Standgaard, H., Phommasack, B., Yun, C.H., Hoang, E.H., 2003. Prevalence of intestinal parasite infections on a national scale among primary schoolchildren in Laos. Parasitol. Res. 91, 267–272.
Sadun, E.H., 1955. Studies on *Opisthorchis viverrini* in Thailand. Am. J. Hyg. 62, 81–115.
Saengsawang, P., Promthet, S., Bradshaw, P., 2013. Infection with *Opisthorchis viverrini* and use of praziquantel among a working-age population in northeast Thailand. Asian Pac. J. Cancer Prev. 14, 2963–2966.
Saengsawang, P., Promthet, S., Bradshaw, P., 2016. Reinfection by *Opisthorchis viverrini* after treatment with praziquantel. Asian Pac. J. Cancer. Prev. 17, 857–862.
Saenna, P., Hurst, C., Echaubard, P., Wilcox, B.A., Sripa, B., 2017. Fish sharing as a risk factor for *Opisthorchis viverrini* infection: evidence from two villages in north-eastern Thailand. Infect. Dis. Poverty 6, 66.

Saijuntha, W., Sithithaworn, P., Wongkham, S., Laha, T., Chilton, N.B., Petney, T.N., Barton, M., Andrews, R.H., 2008. Mitochondrial DNA sequence variation among geographical isolates of *Opisthorchis viverrini* in Thailand and Lao PDR, and phylogenetic relationships with other trematodes. Parasitology 135, 1479–1486.

Saijuntha, W., Sithithaworn, P., Wongkham, S., Laha, T., Pipitgool, V., Tesana, S., Chilton, N.B., Petney, T.N., Andrews, R.H., 2007. Evidence of a species complex within the food-borne trematode *Opisthorchis viverrini* and possible co-evolution with their first intermediate hosts. Int. J. Parasitol. 37, 695–703.

Saiyachak, K., Tongsotsang, S., Saenrueang, T., Moore, M.A., Promthet, S., 2016. Prevalence and factors associated with *Opisthorchis viverrini* infection in Khammouane Province, Lao PDR. Asian Pac. J. Cancer. Prev. 17, 1589–1593.

Saowakontha, S., Pipitgool, V., Pariyanonda, S., Tesana, S., Rojsathaporn, K., Intarakhao, C., 1993. Field trials in the control of *Opisthorchis viverrini* with an integrated programme in endemic areas of northeast Thailand. Parasitology 106 (Pt. 3), 283–288.

Sayasone, S., Mak, T.K., Vanmany, M., Rasphone, O., Vounatsou, P., Utzinger, J., Akkhavong, K., Odermatt, P., 2011. Helminth and intestinal protozoa infections, multiparasitism and risk factors in Champasack Province, Lao People's Democratic Republic. PLoS Neglected Trop. Dis. 5, e1037.

Sayasone, S., Odermatt, P., Phoumindr, N., Vongsaravane, X., Sensombath, V., Phetsouvanh, R., Choulamany, X., Strobel, M., 2007. Epidemiology of *Opisthorchis viverrini* in a rural district of southern Lao PDR. Trans. R. Soc. Trop. Med. Hyg. 101, 40–47.

Sinuon, M., Anantaphruti, M.T., Socheat, D., 2003. Intestinal helminthic infections in schoolchildren in Cambodia. Southeast Asian J. Trop. Med. Public Health 34, 254–258.

Sithithaworn, P., Andrews, R.H., Nguyen, V.D., Wongsaroj, T., Sinuon, M., Odermatt, P., Nawa, Y., Liang, S., Brindley, P.J., Sripa, B., 2012. The current status of opisthorchiasis and clonorchiasis in the Mekong Basin. Parasitol. Int. 61, 10–16.

Sithithaworn, P., Haswell-Elkins, M., 2003. Epidemiology of *Opisthorchis viverrini*. Acta Trop. 88, 187–194.

Sithithaworn, P., Petney, T.N., Andrews, R.H., 2015. What significance do helminths species-complexes have for the prevention, diagnosis and treatment of human infections? Trans. R. Soc. Trop. Med. Hyg. 109, 289–290.

Sithithaworn, P., Pipitgool, V., Srisawangwong, T., Elkins, D.B., Haswell-Elkins, M.R., 1997. Seasonal variation of *Opisthorchis viverrini* infection in cyprinoid fish in northeast Thailand: implications for parasite control and food safety. Bull. World Health Organ. 75, 125–131.

Sithithaworn, P., Tesana, S., Pipitgool, V., Kaewkes, S., Pairojkul, C., Sripa, B., Paupairoj, A., Thaiklar, K., 1991. Relationship between faecal egg count and worm burden of *Opisthorchis viverrini* in human autopsy cases. Parasitology 102 (Pt. 2), 277–281.

Sohn, W.M., Shin, E.H., Yong, T.S., Eom, K.S., Jeong, H.G., Sinuon, M., Socheat, D., Chai, J.Y., 2011. Adult *Opisthorchis viverrini* flukes in humans, Takeo, Cambodia. Emerg. Infect. Dis. 17, 1302–1304.

Sohn, W.M., Yong, T.S., Eom, K.S., Pyo, K.H., Lee, M.Y., Lim, H., Choe, S., Jeong, H.G., Sinuon, M., Socheat, D., Chai, J.Y., 2012. Prevalence of *Opisthorchis viverrini* infection in humans and fish in Kratie Province, Cambodia. Acta Trop. 124, 215–220.

Sornmani, S., Pathammavong, O., Bunnag, T., Impand, P., Intrakhao, C., Thirachantra, S., 1974. An epidemiological survey of human intestinal parasites in Vientiane, Laos. Southeast Asian J. Trop. Med. Publ. Health 5, 541–546.

Sornmani, S., Schelp, F.P., Vivatanasesth, P., Patihatakorn, W., Impand, P., Sitabutra, P., Worasan, P., Preuksaraj, S., 1984a. A pilot project for controlling *O. viverrini* infection in Nong Wai, Northeast Thailand, by applying praziquantel and other measures. Arzneimittelforschung 34, 1231–1234.

Sornmani, S., Vivatanasesth, P., Impand, P., Phatihatakorn, W., Sitabutra, P., Schelp, F.P., 1984b. Infection and re-infection rates of opisthorchiasis in the water resource development area of Nam Pong project, Khon Kaen Province, northeast Thailand. Ann. Trop. Med. Parasitol. 78, 649–656.

Sripa, B., 2018. Tribendimidine: an alternative to praziquantel to treat human liver fluke infection? Lancet Infect. Dis. 18, 124–125.

Sripa, B., Bethony, J.M., Sithithaworn, P., Kaewkes, S., Mairiang, E., Loukas, A., Mulvenna, J., Laha, T., Hotez, P.J., Brindley, P.J., 2011. Opisthorchiasis and Opisthorchis-associated cholangiocarcinoma in Thailand and Laos. Acta Trop. 120 (Suppl. 1), S158–S168.

Sripa, B., Kaewkes, S., Intapan, P.M., Maleewong, W., Brindley, P.J., 2010. Food-borne trematodiases in Southeast Asia epidemiology, pathology, clinical manifestation and control. Adv. Parasitol. 72, 305–350.

Sripa, B., Tangkawattana, S., Laha, T., Kaewkes, S., Mallory, F.F., Smith, J.F., Wilcox, B.A., 2015. Toward integrated opisthorchiasis control in northeast Thailand: the Lawa project. Acta Tropica 141, 361–367.

Sripa, B., Tangkawattana, S., Sangnikul, T., 2017. The Lawa model: a sustainable, integrated opisthorchiasis control program using the EcoHealth approach in the Lawa Lake region of Thailand. Parasitol. Int. 66, 346–354.

Stich, A.H., Biays, S., Odermatt, P., Men, C., Saem, C., Sokha, K., Ly, C.S., Legros, P., Philips, M., Lormand, J.D., Tanner, M., 1999. Foci of *Schistosomiasis mekongi*, Northern Cambodia: II. Distribution of infection and morbidity. Trop. Med. Int. Health 4, 674–685.

Suwannahitatorn, P., Klomjit, S., Naaglor, T., Taamasri, P., Rangsin, R., Leelayoova, S., Mungthin, M., 2013. A follow-up study of *Opisthorchis viverrini* infection after the implementation of control program in a rural community, central Thailand. Parasit Vectors 6, 188.

Suwannatrai, A., Pratumchart, K., Suwannatrai, K., Thinkhamrop, K., Chaiyos, J., Kim, C.S., Suwanweerakamtorn, R., Boonmars, T., Wongsaroj, T., Sripa, B., 2017. Modeling impacts of climate change on the potential distribution of the carcinogenic liver fluke, *Opisthorchis viverrini*, in Thailand. Parasitol. Res. 116, 243–250.

Thaewnongiew, K., Singthong, S., Kutchamart, S., Tangsawad, S., Promthet, S., Sailugkum, S., Wongba, N., 2014. Prevalence and risk factors for *Opisthorchis viverrini* infections in upper Northeast Thailand. Asian Pac. J. Cancer Prev. 15, 6609–6612.

Tomokawa, S., Kobayashi, T., Pongvongsa, T., Nisaygnang, B., Kaneda, E., Honda, S., Moji, K., Boupha, B., 2012. Risk factors for *Opisthorchis viverrini* infection among schoolchildren in Lao PDR. Southeast Asian J. Trop. Med. Publ. Health 43, 574–585.

Touch, S., Komalamisra, C., Radomyos, P., Waikagul, J., 2009. Discovery of *Opisthorchis viverrini* metacercariae in freshwater fish in southern Cambodia. Acta Trop. 111, 108–113.

Upatham, E.S., 1976. Field studies on the bionomics of the free-living stages of St. Lucian *Schistosoma mansoni*. Int. J. Parasitol. 6, 239–245.

Upatham, E.S., 1988. A review of experimental and field research on the human liver fluke, *Opisthorchis viverrini*. J. Sci. Soc. Thailand 14, 245–262.

Upatham, E.S., Viyanant, V., Brockelman, W.Y., Kurathong, S., Lee, P., Kraengraeng, R., 1988. Rate of re-infection by *Opisthorchis viverrini* in an endemic northeast Thai community after chemotherapy. Int. J. Parasitol. 18, 643–649.

Upatham, E.S., Brockelman, W.Y., Viyanant, V., Lee, P., Kaengraeng, R., Prayoonwiwat, B., 1985. Incidence of endemic *Opisthorchis viverrini* infection in a village in northeast Thailand. Am. J. Trop. Med. Hyg. 34, 903–906.

Upatham, E.S., Viyanant, V., 2003. *Opisthorchis viverrini* and opisthorchiasis: a historical review and future perspective. Acta Trop. 88, 171–176.

Upatham, E.S., Viyanant, V., Kurathong, S., Rojborwonwitaya, J., Brockelman, W.Y., Ardsungnoen, S., Lee, P., Vajrasthira, S., 1984. Relationship between prevalence and intensity of *Opisthorchis viverrini* infection, and clinical symptoms and signs in a rural community in north-east Thailand. Bull. World Health Organ. 62, 451–461.

Vannachone, B., Kobayashi, J., Nambanya, S., Manivong, K., Inthakone, S., Sato, Y., 1998. An epidemiological survey on intestinal parasite infection in Khammouane Province, Lao PDR, with special reference to *Strongyloides* infection. Southeast Asian J. Trop. Med. Publ. Health 29, 717–722.

Vo The, D., Waikagul, J., Thanh, B.N., Vo, D.T., Nguyen, D.N., Murrell, K.D., 2014. Endemicity of *Opisthorchis viverrini* liver flukes, Vietnam, 2011–2012. Emerg. Infect. Dis. 20, 152–154.

Vonghachack, Y., Odermatt, P., Taisayyavong, K., Phounsavath, S., Akkhavong, K., Sayasone, S., 2017. Transmission of *Opisthorchis viverrini, Schistosoma mekongi* and soil-transmitted helminthes on the Mekong Islands, southern Lao PDR. Infect. Dis. Poverty 6, 131.

WHO, 2008. Review on the Epidemiological Profile of Helminthiases and Their Control in the Western Pacific Region, 1997–2008.

Wongsaroj, T., Nithikathkulb, C., Rojkitikulc, W., Nakaia, W., Royalb, L., Rammasutc, P., 2014. National survey of helminthiasis in Thailand. Asian Biomed. 8, 779–783.

Wu, X., Lu, Y., Zhou, S., Chen, L., Xu, B., 2016. Impact of climate change on human infectious diseases: empirical evidence and human adaptation. Environ. Int. 86, 14–23.

Wykoff, D.E., Harinasuta, C., Juttijudata, P., Winn, M.M., 1965. *Opisthorchis viverrini* in Thailand—the life cycle and comparison with *O. felineus*. J. Parasitol. 51, 207–214.

Xayaseng, V., Phongluxa, K., van Eeuwijk, P., Akkhavong, K., Odermatt, P., 2013. Raw fish consumption in liver fluke endemic areas in rural southern Laos. Acta Tropica 127, 105–111.

Yong, T.S., Shin, E.H., Chai, J.Y., Sohn, W.M., Eom, K.S., Lee, D.M., Park, K., Jeoung, H.G., Hoang, E.H., Lee, Y.H., Woo, H.J., Lee, J.H., Kang, S.I., Cha, J.K., Lee, K.H., Yoon, C.H., Sinuon, M., Socheat, D., 2012. High prevalence of *Opisthorchis viverrini* infection in a riparian population in Takeo Province, Cambodia. Kor. J. Parasitol. 50, 173–176.

Yoon, H.J., Ki, M., Eom, K., Yong, T.S., Chai, J.Y., Min, D.Y., Rim, H.J., Sohn, W.M., Insisiengmay, B., Phommasack, B., 2014. Risk factors for *Opisthorchis viverrini* and minute intestinal fluke infections in Lao PDR, 2009–2011. Am. J. Trop. Med. Hyg. 91, 384–388.

CHAPTER THREE

Reservoir Animals and Their Roles in Transmission of *Opisthorchis viverrini*

Sirikachorn Tangkawattana*^,§,1 and Prasarn Tangkawattana*

*Faculty of Veterinary Medicine, Khon Kaen University, Khon Kaen, Thailand

§WHO Collaborating Centre for Research and Control of Opisthorchiasis (Southeast Asian Liver Fluke Disease)

[1]Corresponding author: E-mail: sirikach@kku.ac.th

Contents

Abstract

Although any fish-eating mammals could be potential definitive hosts of *Opisthorchis viverrini*, only a few, especially cats and dogs, are actually known reservoir hosts for this parasite. Both animals usually get infected via consuming raw or undercooked contaminated fish, fish dishes or food remains from households. The infected animals sustain parasite egg spread via open environment defecation. Cats are the most important reservoir with higher prevalence rates of *O. viverrini* infection than dogs in endemic areas. Usually *Opisthorchis*-infected animals do not exhibit apparent clinical symptoms or specific abnormalities in laboratory examinations. Pathological findings in these animal reservoirs are basically similar to those seen in humans and experimental animals, namely periductal inflammation, biliary hyperplasia and periductal fibrosis. However, *O. viverrini*-associated cholangiocarcinoma has not yet been reported in

Advances in Parasitology, Volume 101
ISSN 0065-308X
https://doi.org/10.1016/bs.apar.2018.05.003

the reservoir animals at present. Praziquantel is a treatment of choice not only for humans but also for animal reservoirs. Integrated control of opisthorchiasis in animal reservoirs is based on holistic approaches such as EcoHealth/One Health concepts. In fact integrated control of opisthorchiasis in humans in ecosystem has also proved successful, for example, the Lawa model for opisthorchiasis control in the endemic area of Khon Kaen, Thailand. Other feral and wild animals in endemic areas might also be potential reservoirs, and this requires more investigation. In addition, genetic diversity and evolution of the flukes might also influence zoonotic capability.

1. INTRODUCTION

Opisthorchis viverrini was first discovered in an Indian fishing cat, *Prionailurus viverrinus* (Poirier, 1886), with its scientific name incorporating the species name of that host animal. Now a wide range of fish-eating mammals—i.e., dogs, cats, pigs, minks, weasels, civets and house rats—are recognized as final hosts in the life cycle of *O. viverrini* (IARC, 2012). Initially this parasite was believed to just affect animals and hence was not focused on by medical parasitologists. In Thailand, the first two human cases were reported in Chiang Mai in 1911 (Leiper, 1911). Subsequently there have been many epidemiological surveys and research undertaken on its clinical significance to elucidate its life cycle and relationship with community health problems, especially in northern and northeastern Thai provinces (Jongsuksuntigul and Imsomboon, 1998; Sriamporn et al., 2004). Research has revealed several unusual human ailments in connection to the parasitic infection. One intriguing and confirmed finding was its association with a life-threatening cancer, cholangiocarcinoma (CCA) (Sripa and Pairojkul, 2008). This crucial finding has drawn immense multiparty attention in the regions affected—e.g., government, scientists, medical doctors, etc.,—to launch policy and form networks and taskforces to solve these community health problems.

CCA can be induced experimentally in *O. viverrini*-infected Syrian golden hamsters, and many studies have been conducted on the mechanisms of such carcinogenesis. However, *O. viverrini*-infection-associated CCA has never been reported in dogs and cats either with natural or experimental infection. It should be noted that infected animals, similar to humans, rarely exhibit severe clinical signs/symptoms. Thus the life-threatening ability of this parasite in animals is still unclear. To hopefully aid further clarification, this review will analyze information on clinical symptoms, pathology of biliary passages and effective treatments in infected animals.

Extensive studies on various animal species strongly support that dogs and cats, especially cats, are the major sources of natural infection in endemic areas of human opisthorchiasis (Aunpromma et al., 2012). Prevalence of *O. viverrini* infection from our surveys in the super-endemic areas in Khon Kaen province of Thailand always ranges between 0.2% and 0.4% in dogs and 30.9% and 35.5% in cats (Aunpromma et al., 2012, 2016). The high infection prevalence and lack of CCA eruption might encourage veterinary and medical parasitologists to agree that dogs and cats are merely reservoir hosts of *O. viverrini* (WHO, 1995). However, close connection of these animals with humans inevitably brings disease transmission risk, especially via food sharing. In opisthorchiasis endemic areas, rich propagation of the first (*Bithynia* snails) and second (cyprinoid fish) intermediate hosts in wetland environments together with food-consuming practices of the human at-risk groups provide a perfect environment for maintaining the life cycle of *O. viverrini* in the area. People in the endemic area have a tradition of eating raw or undercooked fish dishes made from cyprinoid fish—e.g., *koipla* (minced raw fish salad), *plara* (long-term fermented fish) and *plasom* (short-term fermented fish). They usually feed their animals with food scraps. Moreover, the fish remains that are not disposed of properly in closed containers become the source of infection for feral or stray animals, especially cats. Aunpromma et al. (2012) showed that infection via this mode was higher in cats than in dogs. From the parasite life cycle and the mode of disease transmission, the most effective intervention to establish liver fluke-free communities is to reduce the lack of knowledge about liver fluke transmission via reservoir hosts. However, this information can only be easily provided to owners of tame dogs and cats that are approachable and can be restrained for collection of fecal samples. This review will emphasize the infection sequels of *O. viverrini* in these two carnivores. Investigation on wild animals and sylvatic animals is very rare; however, where available, information regarding the infection in these animals will also be reviewed.

O. viverrini is not the only liver fluke found in animals. To date there are at least 11 small liver fluke species recognized in various fish-eating mammals around the world. Information on their distribution, their animal hosts, major pathological findings and treatments is summarized in Table 1. Among these parasites *O. viverrini* seems to be the most important small liver fluke in animals in the Southeast Asian region. As a common feature all these small liver flukes require the first snail and second fish intermediate hosts and final mammalial hosts to maintain their life cycle. Eggs of the parasite hatch in water (or in the snail host in case of *O. viverrini*) and undergo several

Table 1 Information of Major and Unfamiliar Liver Flukes in Fish-eating Mammals

Spp.	Distribution	Animals	Pathological Findings in Animals	Treatment	References
Opisthorchis viverrini	Cambodia, Central & Southern Vietnam, Laos and Thailand	Cats, civets, dogs, house rats, pigs, minks and weasels	Bile duct proliferation and fibrosis	Fenbendazole (200 mg/kg/day, PO, 3 days) or praziquantel (PQZ) (20–40 mg/kg, PO, once, but 40 mg/kg is much preferable)	Aiello et al. (2016), Anh et al. (2010) Sereerak et al. (2017)
Opisthorchis felineus/ Opisthorchis tenuicollis (Cat/Siberian liver fluke)	Baltic State, Germany, Italy, Kazakhstan, Poland, Russia, Siberia and Ukraine	Cats, dogs, foxes and pigs	Biliary fibrosis (cats, dogs), cholangiohepatitis (cats) and adenocarcinoma of liver or pancreas (cats)	Fenbendazole (200 mg/kg/day, PO, 3 days) or PQZ (20–30 mg/kg, PO, once)	Bowman (2014) Taylor et al. (2007)
Clonorchis sinensis (Oriental liver fluke)	China, Japan, Korea, Northern Vietnam, Russia and Taiwan	Badger sylvatic animals (Badgers, cats, civet cats, dogs, foxes, martens, minks, monkeys, muskrats, pigs, rice rats, and weasels)	Epithelial hyperplasia and fibrosis of the bile duct, cholangiohepatitis and carcinomas (cat)	Dog—PQZ (30 mg/kg/day, PO, 3 days).	Murrell and Pozio (2017) Taylor et al. (2007) Waikagul and Thaekham (2014) Wang et al. (2006)

Platynosomum fastosum *(P. concinnum, P. planicipitus, Dicrocoelium concinnum, D. lanceolatum* var *symmetricum and Concinnum concinnum and P. illiciens)*	Africa (parts), Caribbean Islands, Malaysia, Pacific islands, Puerto Rico, South America and Southeastern USA	Cats	Lizard poisoning from asymptomatic to progressive disease, death due to biliary tract obstruction and hepatic failure, cirrhosis and cholangitis	PQZ (20 mg/kg/day, PO, for 3–5 days, ideally, repeated 12 week later), or nitroscanate (100 mg/kg, PO, once)	Aiello et al. (2016) Basu and Charles (2014) Taylor et al. (2007)
Metorchis albidus	Asia, Europe, North America and USSR (former)	Cats, dogs and foxes	Unnoticed in mild infections. Progressive weakness, complete exhaustion, coma and death in severe cases	Dogs may be attempted with praziquantel (20 mg/kg, PO, once	Aiello et al. (2016) Taylor et al. (2007)
Metorchis conjunctus	Canada, Greenland and North America	Dogs, foxes, raccoons, wolves and dogs	Cholangiohepatitis and periductular fibrosis		Aiello et al. (2016) Waikagul and Thaekham (2014)
Metorchis orientalis	China, Japan and Korea	Cats and dogs	NA		Waikagul and Thaekham (2014)

(*Continued*)

Table 1 Information of Major and Unfamiliar Liver Flukes in Fish-eating Mammals—cont'd

Spp.	Distribution	Animals	Pathological Findings in Animals	Treatment	References
Eurytrema procyonis	Eastern USA	Cats, foxes and raccoons	Injury of pancreatic duct (foxes, raccoons), injury of pancreatic duct, bile duct & gallbladder (cats), pancreatic atrophy and fibrosis and pancreatic periductal fibrosis (foxes, raccoons) and pancreatic epithelial hyperplasia (foxes, raccoons)	No approval dose for fenbendazole, praziquantel, pyrantel and febantel	Aiello et al. (2016) Anderson et al. (1987) Taylor et al. (2007)
Amphimerus pseudofelineus (Cat liver fluke)	Ecuador, and North and Central America	Cats, coyotes, and dogs	Bile duct fibrosis (cats), cirrhosis (cats) and pancreatitis (cats)	Probably praziquantel	Aiello et al. (2016) Bowman et al. (2008) CalvopiCalvopiña (2015) Waikagul and Thaekham (2014)
Pseudamphistomum truncatum	Europe and India	Cats, dogs, and foxes	NA	NA	Taylor et al. (2007)
Parametorchis complexus	North America	Dogs and cats	NA	NA	Taylor et al. (2007)

developmental stages in the first intermediate snail hosts. Then the tailed swimming larvae (cercariae) emerged from snails enter the second intermediate host fish to reach the infective stage (metacercariae) to infect the final hosts. Once metacercariae are ingested by the final hosts, they mature in the biliary apparatus and start to lay eggs. Although the abrasive movement and inflammatory secretion of the mature worms have the potential to injure the liver and biliary apparatus of the infected hosts, reports on clinical symptom in relation to these injuries in animals are very rare. In contrast, *O. viverrini* infection in association with hepatobiliary abnormalities, including fatal cancer, CCA, is long established in humans (Sripa et al., 2012). This information could indicate that animals are likely to be reservoir rather than the final host of the parasites.

2. INFECTION RATES OF *OPISTHORCHIS VIVERRINI* IN ANIMALS

2.1 Infection in Dogs and Cats

Zoonotic nature of *O. viverrini* between cats and humans was described more than 60 years ago by Sadun (1955). Liver fluke eggs in feces of cats and humans were demonstrated to have similar morphology. Coendimicity of *O. viverrini* infection in humans and animals has been reported in Southeast Asian countries such as Thailand and Lao PDR (Sripa et al., 2015; Vonghachack et al., 2017). Although there was no international consensus whether dogs and cats are the only reservoir hosts of this parasite, compilation of several studies on the prevalence in these animals in Thailand provides strong supports for this. Although opisthorchiasis is endemic in the Greater Mekong Subregion, the study of reservoir hosts is limited to Thailand and Lao PDR. The reports and studies on opisthorchiasis prevalence in animal reservoir are summarized in Table 2. In Thailand the highly endemic areas of human infection are located in northeast Thailand, particularly wetland regions, and the prevalence of animal reservoir, particularly cats, is usually high in these areas.

Two recent surveys on *O. viverrini* infection in human and animal reservoirs conducted in a lower northern region of Thailand show a new endemic area of human opisthorchiasis (Pumidonming et al., 2017; Wongsawad et al., 2012), which raises interesting questions. The infection rate for animals in this study area was quite low (0.3% in dogs and 3.3% in cats). Interestingly more than 50% of the residents in this area were the migrants from Northeastern

Table 2 Summary of the Opisthorchiasis Prevalence in Animal Reservoir Reported in Endemic Area of Southeast Asia

Reports	Findings	Study Sites	Prevalence in Animal Reservoir
Thailand			
Sadun (1955)	*Opisthorchis viverrini* infection in cats together with human report. Small-scale survey of cats and dogs in highly endemic and nonendemic area	Endemic area in Northeast Thailand Nonendemic area (Bangkok): cats	Cats: 20% of 20 Dogs: 6.38% 94 Cats: 57.14% (4 of 7)
Wykoff et al. (1965)	Comparative study to differentiate *O. viverrini* from *Opisthorchis felineus*	Northeast Thailand	Cats: 60% Dogs: 40%
Wykoff and Ariyaprakai (1966)	Egg production capacity of *O. viverrini* in cats. Range: 358–3509/cat/day Adult worms residing in the liver: around 99 worms/cat		
Hinz (1980)	Survey of the prevalence of gastrointestinal parasites in dogs	Bangkok, Thailand	Stray dogs: 1.9% of 107
Impand et al. (1983)	Survey of the prevalence in cats and dogs	Khon Kaen and Phitsanulok provinces in Thailand	Cats: 22.6% Dogs: 1.9%
Enes et al. (2010)	Small-scale survey of the prevalence in cats and dogs	Three villages in Khon Kaen	Cats: 36.4% of 22 Dogs: 3.8% of 78
Aunpromma et al. (2012)	Survey of the prevalence in cats and dogs	Chi River Basin and Lawa wetland complex in Khon Kaen	Cats: 35.94% of 217 Dogs: 0.37% of 821

Aunpromma et al. (2016)	Survey of the prevalence in cats and dogs	Ubonrat Dam in Khon Kaen	Cats: 30.92% of 249 Dogs: 0.2% of 1018
Pumidonming et al. (2017)	Survey of the prevalence of gastrointestinal parasites in cats and dogs	Lower Northern Thailand	Cats: 3.3% of 180 Dogs: 3.0% of 197
Neighbouring Countries			
Ditrich et al. (1990)	Survey of the prevalence in human, intermediate hosts and reservoir host	Nam Ngum water reservoir, Lao PDR	Cats: 30% of 27 cats
Giboda et al. (1991)	Survey of the prevalence in human, intermediate hosts and reservoir host	Savannakhet, Lao PDR	Cats: 36% Dogs: 8.8% of 34
Scholz et al. (2003)	Survey of the prevalence in cats	Vientiane, Lao PDR	Cats: 20% of 55
Vonghachack et al. (2017)	Survey of the prevalence in human, intermediate hosts and reservoir host	2 Mekong islands in Southern Lao	Cats: 53.1% Dogs: 25% Pigs: 0.9%

Thailand, who likely brought with them their traditional raw fish-eating practices and cultural lifestyle. If intermediate hosts had already existed in the water body, the fluke could have been introduced either by the migrants or their cats and dogs (Kulsantiwong et al., 2013). However, the low infection rate in animals favours the possibility of introduction of parasites by humans. Another survey in Bangkok was conducted in 2007 with the emphasis on stray or unowned dogs living in temples. All dogs were found to be negative for *O. viverrini* infection, but eight other zoonotic parasite species were found in these dogs (Inpankaew et al., 2007). A further study of fecal samples of 500 dogs and 300 cats from an animal shelter in Nakhon Nayok Province found all of the animals negative for *O. viverrini* (Rojekittikhun et al., 2014). These surveys suggest that *O. viverrini* infection in both the potential animal reservoirs in the central region of Thailand is not a major public health problem. *O. viverrini* infection in animal reservoirs in Lao PDR, a close neighbour of Thailand, has been documented since 1990s (See also Table 2). Similar to Thailand, the surveys suggest that the cat is the important reservoir host because of its higher prevalence. However, there is a report of *O. viverrini* in pigs in southern Lao PDR (Vonghachack et al., 2017). This positive result in pigs suggests that further research is needed on the role of the pig as a reservoir host for this parasite.

Geographical differences might be a potential factor affecting the fecundity of intermediate and definitive hosts which are being influential on the existence of *O. viverrini*. Our study on the severity of infection determined by EPG in dogs and cats along the Chi River Basin and Lawa Lake suggests a positive relationship between infection severity and proximity to the lake (Fig. 1). *O. viverrini* infection prevalence appears to decline from northern to southern latitudes. A study by Kim et al. (2016) showed that the transmission of *O. viverrini* between intermediate hosts and humans is mediated by the ecosystem. Factors such as high salinity and nitrite-nitrogen contaminant levels in the lake correlated with areas with high numbers of *Bithynia* snails and cyprinoid fish. Villages' waste drainage flowed to the southern part of the lake resulting in further contamination with coliform bacteria (Kaewkes et al., 2012). This contaminant-rich water environment is ideal for the *Bithynia* snails to facilitate the parasite spread to that environment. In Thailand the infection rate is highest in the North and Northeast but very rare or negative in the Central and the South regions. Geographical differences are also noticed in other Southeast Asian countries. The infection rate in the vicinal area of Mekong River is higher than in the lower parts of the southward countries. Three

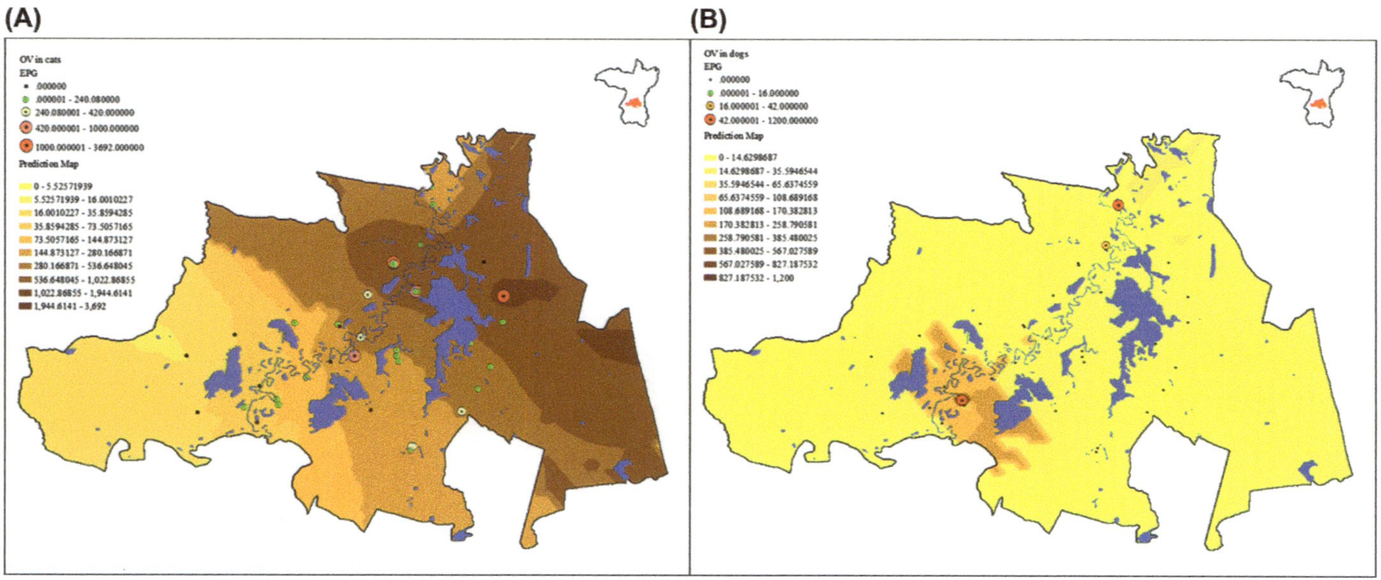

Figure 1 Geographical distribution of cats (A) and dogs (B) with different EPG in Khon Kaen, Thailand. Distribution of intensity of EPG in cats is relative to proximity to the water body.

surveys on *O. viverrini* infection in dogs and cats in Malaysia during 2014–15 were all found to have negative results (Ngui et al., 2014; Tun et al., 2015).

2.2 Infection in Sylvatic Animals

Sylvatic or wild animals and synanthropic animals that live around residential or household areas could also play very important roles as reservoir hosts for other liver fluke species, such as *Clonorchis sinensis* and *O. felineus* (Murrell and Pozio, 2017). However, their role in *O. viverrini* transmission is unclear to date. Many epidemiological studies suggest that wild animals could sustain the life cycle of the liver fluke in the absence of domestic cycle (Murrell and Pozio, 2017). So far at least four families of fish-eating mammals—i.e., *Felidae, Herpestidae, Mustelidae* and *Viverridae*—were indicated as possible reservoir hosts in the Mekong Region (Petney et al., 2013). These include fishing cat (*Prionailurus viverrinus*), leopard cat (*Prionailurus bengalensis*), crab-eating mongoose (*Herpestes urva*), Eurasian otter (*Lutra lutra*), Oriental small-clawed otter (*Aonyx cinerea*), smooth otter (*Lutrogale perspicillata*), civet or large Indian civet (*Viverra zibetha*) and binturong (*Arctictis binturong*). In Thailand we have confirmed *O. viverrini* only in fecal drops of fishing cats from the southern part of Thailand (unpublished data). More studies are crucial to elucidate the role of wildlife as reservoir hosts.

It has been observed that the long-tailed macaque (*Macaca fascicularis*) in Sumatra and Kalimantan catches and eats fish (Stewart et al., 2008). Thus it is possible that this or other macaques can be a potential reservoir for *O. viverrini*, although the survey of intestinal parasites by Wenz-Mucke et al. (2013) did not show any evidence of *O. viverrini* infection in this animal. Recently we conducted a preliminary study on the possibility of liver fluke infection among a colony of about 400 long-tailed macaques in Kosumpee Sanctuary, Kosumpisai, Mahasarakham Province, Thailand (unpublished data). Although geography of this sanctuary is suitable for the propagation of the liver fluke, our preliminary results showed they were all negative. A trial to check whether the macaques in this area eat fish or not was also conducted. We found that they ignored fresh fish being offered. In the meantime, we conclude that liver fluke infection will remain zero in this colony unless they change their eating preferences.

2.3 Avian Liver Fluke

In 2014 Dao et al. discovered *O. viverrini*-like fluke in ducks (*Anas platyrhynchos*) in Binh Dinh Province, Central Vietnam and proposed that avian species could be a nonmammalian reservoir host. However, these worms had some morphological differences from *O. viverrini* isolated from humans, cats and hamsters, and Nawa et al. (2015) claimed that this liver fluke from ducks is highly likely to be *O. parageminus*. The prevalence of this parasite in Binh Dinh Province was 34.3% (Dao et al., 2016a), and the ducks in this area eat raw small wild-caught freshwater fish (*Carassius auratus*) having liver fluke metacercariae (Dao et al., 2016b). They still insist this parasite as '*O. viverrini*, duck genotype'. Recently this duck-derived liver fluke *Opisthorchis* sp. BD2013 was differentiated from *O. viverrini* by molecular phylogenetic analysis (Dao et al., 2017). Whether this is the synonym of *O. parageminus* or this new genotype is derived by 'sympatric speciation' (Fitzpatrick et al., 2008) should be clarified in future using population genetic methods.

3. CAT AS THE MOST IMPORTANT RESERVOIR HOST OF *OPISTHORCHIS VIVERRINI*

O. viverrini infection occurs not only in the fish-eating mammals but also in some laboratory animals such as hamsters, gerbils, guinea pigs, rabbits, mice and rats by orally feeding certain number of metacercariae (IARC, 2012; Wykoff et al., 1965). *O. viverrini* can maturate in hamsters, gerbils, rabbits and guinea pigs, whereas it cannot reach maturity in mice and rats (IARC, 2012; Wykoff et al., 1965), and the worm will be expelled from the host after arrival at the liver. This indicates that mice and rats should not be classified as the final host of *O. viverrini*. So far only dogs and cats are officially recognized as reservoir hosts (WHO, 1995) because they are the habitat for the infective agent to survive, mature, reproduce and be infective to humans (Ashford, 2003). In addition, reservoir host can get either single or repeated infections without or with minimal clinical symptoms (Aguirre et al., 2012).

Regardless of the scale of the survey, almost all studies in Thailand and other countries showed that the *O. viverrini* infection rates in cats were much higher than those in dogs. The infection record in other animal species was very rare. Ranges of the infection prevalence in cats were 30.9%–36.4% in Khon Kaen of Thailand (Enes et al., 2010, Aunpromma et al., 2012, 2016), 3%–60% in other parts of Thailand (Sadun, 1955, Wykoff et al., 1965, Impand et al., 1983, Enes et al., 2010, Aunpromma et al., 2012, 2016, Pumidonming et al., 2017) and 20%–35% in Lao PDR (Ditrich et al., 1990; Giboda et al., 1991; Scholz et al., 2003; Vonghachack et al., 2017). In comparison, the prevalence in dogs compiling from different surveys had a range of 0.2%–3.8% in Thailand (Impand et al., 1983; Enes et al., 2010; Aunpromma et al., 2012, 2016; Wongsawad et al., 2012; Pumidonming et al., 2017) and 8.8%–25% in Lao PDR (Vonghachack et al., 2017). These data clearly indicate that cat should be considered the most important reservoir host of *O. viverrini*.

Adult liver flukes residing in biliary apparatus release eggs into bile which are further conveyed into the intestine. Wykoff and Ariyaprakai (1966) reported that egg production rate of *O. viverrini* in cats ranged from 358 to 3509 eggs/cat/day. Stool examination for fluke eggs has been the gold standard for the diagnosis of opisthorchiasis. Although number of eggs being released from the adult worms into bile might be different among infected individuals, EPG has been used for grading infection severity into four different grades—i.e., mild (EPG < 50), moderate (EPG $= 50-500$), heavy (EPG $= 500-1500$) and very heavy (EPG > 1500) (Aunpromma et al., 2012). In terms of egg production of *O. viverrini*, cats have been found to be more productive than human, hamsters and rabbits (Wykoff et al., 1965; Wykoff and Ariyaprakai, 1966). In a necropsy of an infected cat with EPG of more than 6000, the authors harvested about 300 adult worms from just one half of its liver (unpublished data). However, average EPG in cats of previous studies usually falls in the mild or moderate severity (Ditrich et al., 1990, Aunpromma et al., 2012, Sereerak et al., 2017). Heavily infected cats with average EPG of 615 were recently detected in a high endemic area of Khon Kaen, Thailand (Aunpromma et al., 2016). A cat in this survey was found to have EPG of 17,504. This survey covered all aspects in relation to risk of infection, and it was clear that age and eating habit of the infected cats were closely associated with the infection severity. The youngest cats found infected with the parasite were 3- to 6-month old (Enes et al., 2010), but infection was usually found in cats older than 3 years rather than in cats younger than 1 year (Aunpromma et al., 2016).

Lacking of the infection status in stray cats is a problem for current situation of this reservoir. Some adult cats become feral and aggressive in their hunting behaviour (Kim et al., 2017) and might live on catching fish at the river or open water body. However, our data (Aunpromma et al., 2012) and recent ongoing participatory epidemiological work indicate that they always come back to households to sneak food scraps or raw fish remaining from cooking. Because fish in these high endemic areas usually contain metacercariae, eating them raw is a high infection risk behaviour. Chai et al. (2013) also indicated that the chance of exposing and consuming contaminated raw fish for feral or stray cats is much higher than the owned cats. Regardless of infection severity, the infected cats are usually clinically normal. Our clinical and ultrasonographic examination of naturally infected cats from high endemic areas of Khon Kaen, Thailand, confirm this (unpublished data). Thus stray cats could also play an important role as a reservoir host for *O. viverrini*. Stool collection from the unowned or feral cats is extremely difficult because the animals will not allow the unfamiliar collectors to approach them. As there are typically cats missing from the surveys, the real infection prevalence in cats is always an underestimation.

4. CLINICOPATHOLOGICAL FEATURES OF *OPISTHORCHIS VIVERRINI*-INFECTED CATS

Clinical and haematological examinations of dogs and cats with natural infection of *O. viverrini* from highly endemic areas of human opisthorchiasis confirm that infected animals did not show any apparent clinical symptoms. Usually their clinical and haematological profiles were similar to those of healthy animals (Aunpromma et al., 2012; Sereerak et al., 2017). Blood and biochemical profiles of cats with *O. viverrini* infection could not be distinguished from those of clinically normal cats. Eosinophilia that usually occurs in parasitic infestation did not happen in these cases. In addition, hepatic enzymes, which should be altered if hepatic parenchyma is injured, were not different from those of the normal uninfected cats (Aunpromma et al., 2012; Sereerak et al., 2017). In these studies only a few with very heavy infection showed signs of lethargy and mild diarrhoea, which could be a complication from other causes, such as intestinal nematodes or other ailments. Thus it is not possible to use only physical examination for the diagnosis of animal opisthorchiasis. This may be different for infection with other liver flukes, such as *Platynosomum fastosum*. Infection severity and clinical symptoms of feline platynosomiasis caused by *P. fastosum* (Basu and Charles, 2014) varied from asymptomatic to progressive. Bile duct obstruction and liver failure occurring in the late stage of the disease could be the causes of death of the animals. Significant increases in alanine transaminase or serum glutamic pyruvic transaminase were usually detected in most cases with hepatic damage (Basu and Charles, 2014; Ramos et al., 2017). However, these enzymes could also be elevated because of injured skeletal and heart muscles or hyperthyroidism (Aiello et al., 2016). Thus the use of this enzymatic profile as a single tool for the interpretation of disease should be avoided.

Ultrasonography is a powerful and noninvasive diagnostic tool for many liver diseases in humans and animals including cats. Ultrasonogram of the *Platynosomum* spp.-infected cats can demonstrate sonographic changes of the bile duct and gallbladder (Salomao et al., 2005). Wall thickening and distention of the gallbladder with hyperechoicity and anechoic content, as well as distention and torsion of bile ducts, were also revealed. Ultrasonography was applied also to the naturally *O. viverrini*-infected cats. Accordingly, sonographic changes of hepatic parenchyma and gallbladder were obtained (Fig. 2). This changing pattern was similar to that in human opisthorchiasis (Mairiang et al., 1992, 2012) and MRI image in laboratory animals (Hanpanich et al., 2013, 2017) which included periductal fibrosis and bile duct dilatation. These alterations were putatively explained as the results of liver fluke and host responses (Sripa and Pairojkul, 2008). In addition, investigation using MRI and CT scan on feline liver changes also yielded similar results to sonography (unpublished data).

Pathological changes of liver fluke infection caused by *O. viverrini* and *C. sinensis* in humans and animal models have been studied for decades. IARC (1994, 2012) classified both parasites as group I biological agents carcinogenic to humans. Major histopathological features expressed by these worms are in line with inflammatory process and cell proliferation. Eosinophil population is predominant in the early phase of inflammation. Declination of eosinophil is reversed with an increase of mononuclear cells—e.g., lymphocytes, plasma cells and macrophages. Fibrosis or granuloma formation can also occur from the penetration of the parasites. Usually slight changes in gross or necropsy findings in opisthorchiatic animals at chronic phase of infection are observed. Hepatic parenchyma is still normal or shows rough surface. The gallbladder and major bile ducts are thickened and translucent. Bile fluid shows increased viscosity and heterogeneity. Adult worms mostly reside in the large bile duct and gallbladder where fibrosis occurs (Fig. 2). Histopathological changes of biliary cells in the *O. viverrini*-infected cases include epithelial desquamation, epithelial and adenomatous hyperplasia, goblet cell metaplasia and periductal fibrosis. Eruption of CCA can only be found in humans. Moreover, induction of CCA in Syrian hamster, an animal model for opisthorchiasis, always requires carcinogen enhancement, such as dimethylnitrosamine together with the *O. viverrini* infection (IARC, 1994, 2012, Sripa et al., 2003, 2007). It is not common for liver fluke-associated CCA to develop in the naturally infected animals. Although there were a few reports of CCA in cats caused by other liver flukes, such as *Platynosomum fastosum*

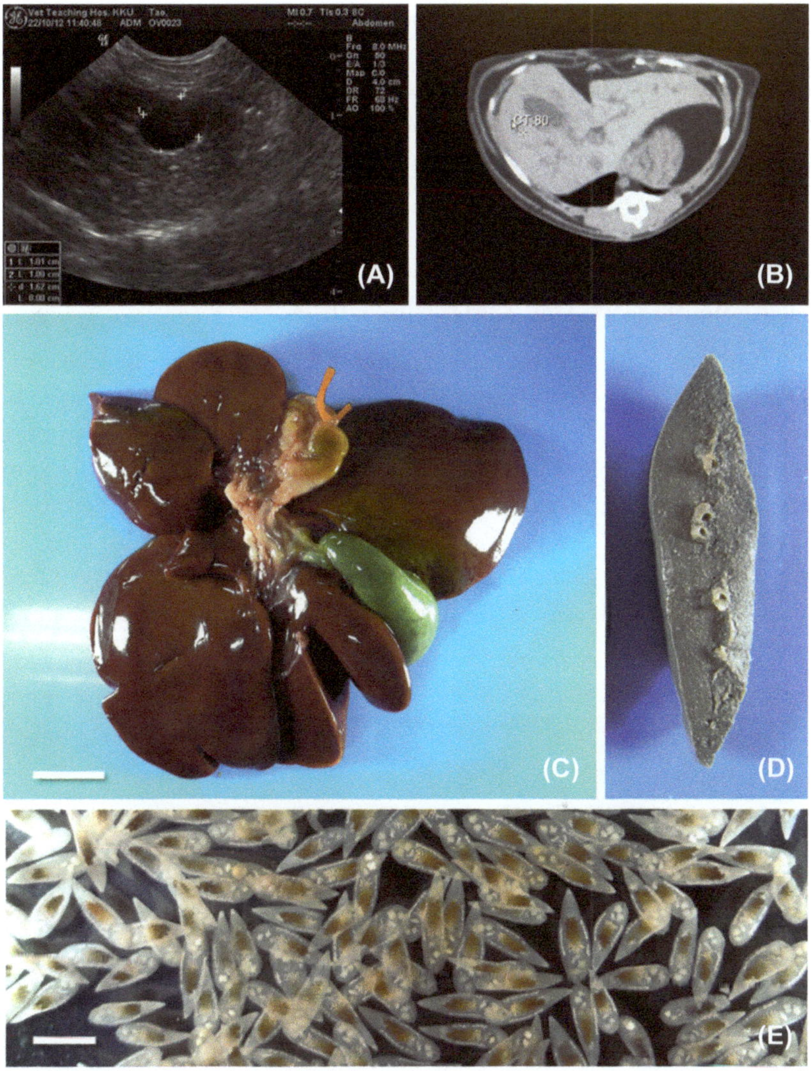

Figure 2 Pathological findings of *Opisthorchis viverrini*-infected liver of cats. (A) Ultrasonogram, (B) CT image, (C) Liver of the infected cat showed slight enlargement and distended gallbladder with thickening bile duct (bar = 15 mm), (D) Cross section of formalin-fixed cat liver with adult flukes residing inside bile ducts, and (E) Adult worms from biliary tract (bar = 5 mm).

(Andrade et al., 2012; Santos et al., 1981), *O. viverrini*-associated CCA in animal reservoirs has never been reported at this time. There are at least three presumptions to explain why CCA does not develop in the infected animals. First, lifespan of the infected animals is not long enough for CCA development. Usually CCA development in human takes decades after the infection. Second, host immune response, particularly inflammation, is not aggressive or highly active enough to trigger the eruption of CCA. Finally, species specificity is a determinant of CCA development. Detailed studies on these three presumptions are required.

5. ROLE OF RESERVOIR HOSTS IN SUSTAINING PARASITE IN THE ENDEMIC AREA

Cats and dogs are the main animal reservoirs in the life cycle of *O. viverrini* (Fig. 3). Recently using a disease model of opisthorchiasis, Bürli et al. (2018) indicated that cats were more influential on the worm burden in humans than in dogs. In rural endemic area in northeastern Thailand most of the reservoir animals, particularly cats, are unowned and allowed to roam freely. They must search for food by themselves (Aunpromma et al., 2012; Petney et al., 2013). This independency would increase the possibility of infection of many diseases, including liver fluke infection, from the contaminated environment.

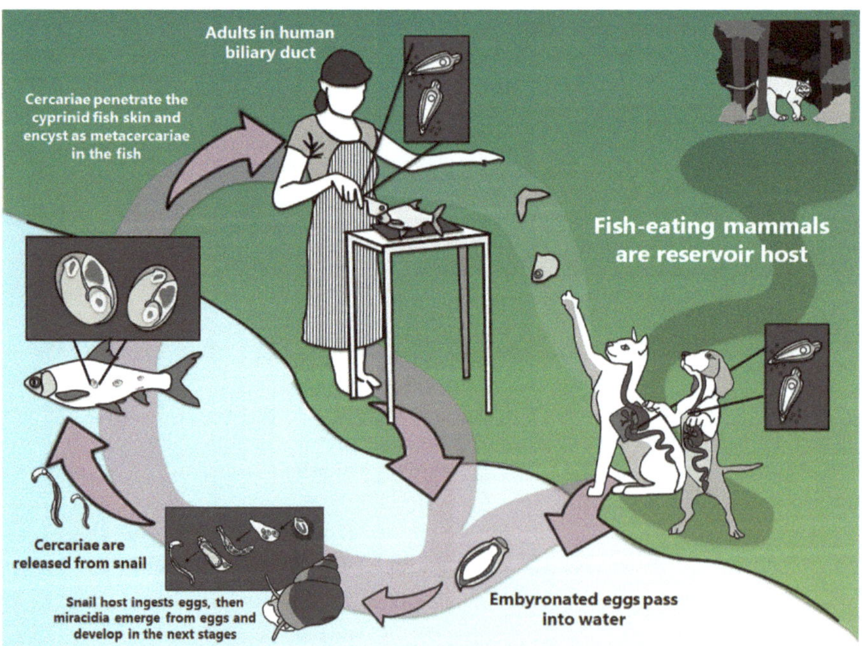

Figure 3 Cats and dogs as animal reservoirs in the life cycle of *Opisthorchis viverrini*.

The high endemic areas of opisthorchiasis usually have fertile environment—i.e., lowland, wetland and flooding—necessary for the reproduction and propagation of the intermediate hosts and the parasites as well (Sithithaworn et al., 2012; Wang, 2012). All these factors are essentially required for the completion of the life cycle of the parasite. Most animal reservoirs can roam freely either near or far from the household areas. Although they might get infected from eating infected fish that they caught at the river or water body, they have much higher chance of infection if they sneak in the house and eat undercooked scrap fish or fish remains in uncovered household garbage bins. Free-roaming animals usually defecate on land, and parasite eggs in faeces are flushed into the water body either by rain or flooding (Aunpromma et al., 2012). The eggs are then ingested by *Bithynia* snails and contribute to the maintenance of the life cycle of *O. viverrini*. As mentioned previously, Bürli et al. (2018) used a mathematical model to simulate the transmission dynamic of *O. viverrini* in two islands in Lao PDR to compare the impact of the presence/absence of the reservoir hosts (cats and dogs). They concluded that human host alone is sufficient to maintain the life cycle of parasites and suggested that disease transmission can be interrupted without treating or controlling the reservoir hosts. Petney et al. (2013) also stipulated that the animal reservoirs are less important than human host. In contrast, Murrell and Pozio (2017) stated that wild animals could sustain the liver fluke even in the absence of domestic cycle. Because mathematical models are created with limited assumptions—e.g., ignorance of the significance of heterogeneities, seasonal and age effects or geographical conditions—more improved mathematical models integrating all those variances are necessary for the actual evaluation of the significance of reservoir animals for the maintenance of the life cycle and spreading of *O. viverrini* to the environment.

6. TREATMENT AND CONTROL OF OPISTHORCHIASIS FOR ANIMAL RESERVOIRS

Effective disease control of opisthorchiasis requires a holistic approach of all stakeholders from various disciplines. System thinking with EcoHealth/One Health concept, which gathers all stakeholders to participate equally in the control program, could be a solution of choice to be introduced and applied in the endemic area. Recently a One Health approach was recommended for programs for controlling food-borne trematodiases including opisthorchiasis (WPRO, 2017). Understanding of the nature of disease and the transmission among a range of hosts would help targeting the key players and lead to the effective control. The model was recently selected to be a WHO showcase in food-borne parasitic disease control (Sripa et al., 2015, 2017; WHO, 2011). The concept of One Health, which connects human health, animal health and ecosystem, seems to match with the control of this parasitic disease. Practically the treatment and control of animal reservoir should be based on routine animal healthcare (feeding, deworming and vaccination), human personal hygiene (knowledge, attitude and practice) and waste management.

Praziquantel (PZQ) is a drug of choice for the treatment of opisthorchiasis in the reservoir hosts as well as humans. The recommended dose for small liver flukes such as *Opisthorchis* spp. is 20 mg/kg, per os (PO), once, and for *Platynosomum* spp. is 20 mg/kg/day, PO, for 3−5 days, and it is repeated once 12 weeks later (Aiello et al., 2016). Recently a dose of 20 mg/kg PZQ was shown to be insufficient to obtain complete cure of heavy infection in cats. Accordingly a single dose of 40 mg/kg, PO, once, was proven to be effective and safe for heavily infected cats (Sereerak et al., 2017). Although there is no systematic research to determine the dosage regimen for dogs, treatment dose at 40 mg/kg could be applied from the study of *Clonorchis* infection in dogs (Anh et al., 2010). The risk groups of both humans and animals should have regular deworming program every 6 months to prevent the new infection and reinfection.

The Centers for Disease Control and Prevention (CDC) and American Association of Veterinary Parasitologists (AAVP) have recommended deworming the common intestinal parasites in cats and dogs since they are young. A deworming program for puppies and kittens should be given at the age of 2, 4, 8, 12, 16, 24 weeks and 12 months. The adults should be dewormed every 6 months (Division of Parasitic Diseases, CDC/NCID/DPD, 1995). In Thailand or other *O. viverrini*-endemic countries routine deworming of cats and dogs for liver fluke has not been programmed. Only therapeutic doses are suggested for the infected animals. Therefore the therapeutic dose should be applied for routine deworming of liver flukes for dogs and cats in the high endemic areas.

Because many omnivorous mammals can be the final hosts of *O. viverrini*, anthelmintic treatment of all reservoir animals in the endemic areas is practically impossible. The most effective and practical approach to control liver fluke infections in reservoir hosts is to educate the community. Giving knowledge to the residents in the endemic areas about zoonotic diseases, nature of parasitic diseases and its transmission in animals is very crucial. Health officers, veterinarians and other stakeholders should promote and raise public awareness on this issue. The attitude of pet owners towards pet raising and healthcare for their pets needs to be improved. The owners should not feed or allow their pets to freely roam to scavenge raw scrap food or fish remains because it is a high-risk behaviour. Furthermore, improvement of attitude about animal waste management and awareness about zoonotic infection will help the involved parties to improve healthcare for their animals. Matos et al. (2015) also raised the problem of ineffective treatment because of irregular treatment of companion animals with antiparasitic drugs. Integrated approaches for effective and sustainable management are necessary in control programs for animal reservoirs and humans. In regions where aquaculture is expanding, control programs should be extended to the intermediate and reservoir hosts (Dung et al., 2014).

Liver fluke crucially requires suitable aquatic environments to facilitate egg transmission from definitive hosts to specific snails and cyprinoid fish intermediate hosts. Endemicity of *O. viverrini* in wetland is usually determined by the fecundity of these factors required for the completion of the life cycle of the parasite (Aunpromma et al., 2012; Otake Sato et al., 2017; Vonghachack et al., 2017). Because parasite eggs are released into environment via the stool of human and animal reservoir hosts, the lack of good sanitation and effective waste management are major causes of parasite egg dispersal. Latrine installation in every household and rice field could help prevent transmission from humans to environment.

Animal waste with high contents of phosphorus, nitrogen and other nutrient can pollute the environment through promoting '*algae bloom*' in water bodies. In addition, it can convey other pathogens such as bacteria and parasites that can be transmitted to humans or other animals either directly or indirectly (CTAHR, 2000). Because reservoir animals usually defecate into environment directly, disposal of the animal manure is necessary. Although animal wastes can be used as fertilizer for plants, pretreatment or degradation to eliminate the parasite is required. If an incineration system is available, manure in degradable bags can be incinerated before disposal in the landfill. Higher technology, such as combined system of an up-flow anaerobic hybrid reactor and a down-flow hanging sponge system, could be applied to eliminate the parasite in the waste (Tawfik et al., 2015). However, practical and economic management for all endemic areas to handle this problem still requires more development. Overpopulation of cats and dogs and stray or feral animals is a major problem in many cities or countries. It affects by spreading not only the liver fluke but also other pathogens, particularly zoonotic pathogens. Control of animal population requires practical policy for solving the problems in all levels, such as community, regional or national level. Policy and regulation should focus on pet owners' responsibility and control of stray animals.

7. CONCLUSION AND FUTURE DIRECTION

Opisthorchiasis due to *O. viverrini* infection is a public health problem in Southeast Asia. Chronically infected patients can develop fatal CCA in the late stage of infection. Infected animals become reservoirs for the spread of the parasite to the environment. As a fish-eating mammal, cats are considered the major animal reservoir in endemic areas. Completion of the life cycle of the parasite requires environment suitable for reproduction and propagation of the intermediate hosts (*Bithynia* snails and cyprinoid fish) and transmission of parasite eggs from the definitive human host and reservoir animals (cats and dogs) to the water body. All these factors are plentiful in wetlands. Control and treatment of opisthorchiasis in human patients can be handled effectively. Although many epidemiological surveys of *O. viverrini* infection in many endemic areas have been conducted, the real infection status in reservoir animals, domestic, feral or wildlife animals, is still far from clear. A practical control measure for the reservoir hosts still requires further development.

ACKNOWLEDGEMENTS

This study was supported by the Higher Education Research Promotion and National Research University Project of Thailand, Office of the Higher Education Commission, through the Center of Excellence in Specific Health Problems in Greater Mekong Subregion Cluster (SHeP-GMS, Grant No.H-2553-M-24), Khon Kaen University and the IDRC Project (Component Number: 108100-001), Annual Seed Grant Year 2012 from Faculty of Veterinary Medicine, Khon Kaen University Annual Grant (Year 2/2010) and Thailand Research Fund Senior Research Scholar (Interdisciplinary Research on Human Liver Flukes, PI- Prof. Banchob Sripa, RTA5680006).

The authors would like to thank Dr. Apiporn Suwannatrai for the prediction map construction, Dr. Tomas Leon for explanation of the mathematical model, Dr. Watcharapol Suyapoh for the diagram describing animal reservoir in the parasite's life cycle and Dr. Nathathai Wannachalerm, Dr. Piya Sereerak and Dr. Songkiad Upontain for technical assistances. The authors would also like to acknowledge Prof. Yukifumi Nawa for editing the manuscript via Publication Clinic of KKU, Thailand and Assoc Prof. John F. Smith for English editing.

REFERENCES

Aguirre, A.A., Ostfeld, R., Daszak, P., 2012. New Directions in Conservation Medicine: Applied Cases of Ecological Health. Oxford University Press.

Aiello, S.E., Moses, A.A., Allen, D.G., 2016. The Merck Veterinary Manual, eleventh ed. Merck & Co., Inc., Kenilworth, NJ.

Anderson, W.I., Georgi, M.E., Car, B.D., 1987. Pancreatic atrophy and fibrosis associated with *Eurytrema procyonis* in a domestic cat. Vet. Rec. 120, 235–236.

Andrade, R.L., Dantas, A.F., Pimentel, L.A., Galiza, G.J., Carvalho, F.K., Costa, V.M., et al., 2012. *Platynosomum fastosum*-induced cholangiocarcinomas in cats. Vet. Parasitol. 190, 277–280.

Anh, N.T., Madsen, H., Thanh, D.T., Dalsgaard, A., Murrell, K.D., 2010. Field trial of praziquantel for control of fishborne zoonotic trematodes in reservoir hosts in Vietnam. Vet. Parasitol. 174, 348–350.

Ashford, R.W., 2003. When is a reservoir not a reservoir? Emerg. Infect. Dis. 9, 1495–1496.

Aunpromma, S., Kanjampa, P., Papirom, P., Tangkawattana, S., Tangkawattana, P., Tesana, S., et al., 2016. Prevalence and risk factors for *Opisthorchis viverrini* infection among cats and dogs in six districts surrounding the Ubonratana Dam, an endemic area for human opisthorchiasis in Northeastern Thailand. Southeast Asian J. Trop. Med. Publ. Health 47, 1153–1159.

Aunpromma, S., Tangkawattana, P., Papirom, P., Kanjampa, P., Tesana, S., Sripa, B., et al., 2012. High prevalence of *Opisthorchis viverrini* infection in reservoir hosts in four districts of Khon Kaen Province, an opisthorchiasis endemic area of Thailand. Parasitol. Int. 61, 60–64.

Basu, A.K., Charles, R.A., 2014. A review of the cat liver fluke *Platynosomum fastosum* Kossack, 1910 (Trematoda: *Dicrocoeliidae*). Vet. Parasitol. 200, 1–7.

Bowman, D.D., 2014. Georgis' Parasitology for Veterinarians. 10th ed. Elsevier Science Health Science, St. Louise.

Bowman, D.D., Hendrix, C.M., Lindsay, D.S., Barr, S.C., 2008. Feline Clinical Parasitology. John Wiley & Sons.

Bürli, C., Harbrecht, H., Odermatt, P., Sayasone, S., Chitnis, N., 2018. Mathematical analysis of the transmission dynamics of the liver fluke, *Opisthorchis viverrini*. J. Theor. Biol. 439, 181–194.

Calvopiña, M., Cevallos, W., Atherton, R., Saunders, M., Small, A., Kumazawa, H., et al., 2015. High prevalence of the liver fluke *Amphimerus* sp. in domestic cats and dogs in an area for human amphimeriasis in Ecuador. PLoS Neglected Trop. Dis. 9, e0003526.

Chai, J.Y., Bahk, Y.Y., Sohn, W.M., 2013. Trematodes recovered in the small intestine of stray cats in the Republic of Korea. Kor. J. Parasitol. 51, 99–106.

CTAHR, 2000. Pet Waste Management. Hawaii's Pollution Prevention Information. HAPPI-Home 16. https://www.ctahr.hawaii.edu/oc/freepubs/pdf/hh-16.pdf.

Dao, T.H., Nguyen, T.G., Victor, B., Gabriël, S., Dorny, P., 2014. *Opisthorchis viverrini*-like liver fluke in birds from Vietnam: morphological variability and rDNA/mtDNA sequence confirmation. J. Helminthol. 88, 441–446.

Dao, T.T., Abatih, E.N., Nguyen, T.T., Tran, H.T., Gabriël, S., Smit, S., 2016a. Prevalence of *Opisthorchis viverrini*-like fluke infection in ducks in Binh Dinh province, Central Vietnam. Kor. J. Parasitol. 54, 357–361.

Dao, T.T., Bui, T.V., Abatih, E.N., Gabriël, S., Nguyen, T.T., Huynh, Q.H., 2016b. *Opisthorchis viverrini* infections and associated risk factors in a lowland area of Binh Dinh Province, Central Vietnam. Acta Trop. 157, 151–157.

Dao, T.T.H., Nguyen, T.T.G., Gabriël, S., Bui, K.L., Dorny, P., Le, T.H., 2017. Updated molecular phylogenetic data for *Opisthorchis* spp. (Trematoda: *Opisthorchioidea*) from ducks in Vietnam. Parasit. Vectors 10, 575.

Ditrich, O., Scholz, T., Giboda, M., 1990. Occurrence of some medically important flukes (Trematoda: *Opisthorchiidae* and *Heterophyidae*) in Nam Ngum water reservoir, Laos. Southeast Asian J. Trop. Med. Publ. Health 21, 482–488.

Division of Parasitic Diseases, CDC/NCID/DPD, 1995. Recommendations for Veterinarians; How to Prevent Transmission of Intestinal Roundworms from Pets to People. https://wonder.cdc.gov/wonder/prevguid/p0000424/p0000424.asp#head006000000000000.

Dung, V.T., Waikagul, J., Thanh, B.N., Vo, D.T., Nguyen, D.N., Murrell, K.D., 2014. Endemicity of *Opisthorchis viverrini* liver flukes, Vietnam, 2011–2012. Emerg. Infect. Dis. 20, 152–154.

Enes, J.E., Wages, A.J., Malone, J.B., Tesana, S., 2010. Prevalence of *Opisthorchis viverrini* infection in the canine and feline hosts in three villages, Khon Kaen Province, northeastern Thailand. Southeast Asian J. Trop. Med. Publ. Health 41, 36–42.

Fitzpatrick, B.M., Fordyce, J.A., Gavrilets, S., 2008. What, if anything, is sympatric speciation? J. Evol. Biol. 21, 1452–1459.

Giboda, M., Ditrich, O., Scholz, T., Viengsay, T., Bouaphanh, S., 1991. Current status of food-borne parasitic zoonoses in Laos. Southeast Asian J. Trop. Med. Publ. Health 22 (Suppl.), 56–61.

Hanpanich, P., Laha, T., Sripa, B., Mairiang, E., Sereerak, P., Upontain, S., et al., 2017. Decreased risk of cholangiocarcinogenesis following repeated cycles of *Opisthorchis viverrini* infection-praziquantel treatment: magnetic Resonance Imaging (MRI) and histopathological study in a hamster model. Parasitol. Int. 66, 464–470.

Hanpanich, P., Pinlaor, S., Charoensuk, L., Yongvanit, P., Thomas, C., Kothan, S., et al., 2013. MRI and $^{(1)}$H MRS evaluation for the serial bile duct changes in hamsters after infection with *Opisthorchis viverrini*. Magn. Reson. Imaging 31, 1418–1425.

Hinz, E., 1980. Intestinal helminths in Bangkok stray dogs and their role in public health (author's tranls). Zentralbl. Bakteriol. Mikrobiol. Hyg. B 171, 79–85 (German, English abstract).

IARC, 1994. Schistosome. Liver Flukes and *Helicobacter pylori*. In: IARC Monographs Carcinogenic Risks to Human, vol. 61. International Agency for Research on Cancer, Lyon.

IARC, 2012. IARC Monographs on the Evaluation of Carcinogenic Risks to Humans, vol. 100B. International Agency for Research on Cancer, Lyon.

Impand, P., Thirachandra, S., Bunnag, T., 1983. Helminth faunas of rats and domestic animals and their zoonotic potential role in north and northeast Thailand. J. Parasit. Trop. Med. Assoc. Thai. 6, 105–116.

Inpankaew, T., Traub, R., Thompson, R.C., Sukthana, Y., 2007. Canine parasitic zoonoses in Bangkok temples. Southeast Asian J. Trop. Med. Publ. Health 38, 247–255.

Jongsuksuntigul, P., Imsomboon, T., 1998. Epidemiology of opisthorchiasis and national control program in Thailand. Southeast Asian J. Trop. Med. Publ. Health 29, 327–332.

Kaewkes, W., Kaewkes, S., Tesana, S., Laha, T., Sripa, B., 2012. Fecal bacterial contamination in natural water reservoirs as an indicator of seasonal infection by *Opisthorchis viverrini* in snail intermediate hosts. Parasitol. Int. 61, 49–51.

Kim, C.S., Echaubard, P., Suwannatrai, A., Kaewkes, S., Wilcox, B.A., Sripa, B., 2016. Seasonal and spatial Environmental influence on *Opisthorchis viverrini* intermediate hosts, abundance, and distribution: insights on transmission dynamics and sustainable Control. PLoS Neglected Trop. Dis. 10, e0005121.

Kim, S., Cho, Y.S., Bhak, J., O'Brian, S.J., Yeo, J.H., 2017. Perspectives provided by leopard and other cat genomes: how diet determined the evolutionary history of carnivores, omnivores, and herbivores. BMB Rep. 50, 3–4.

Kulsantiwong, J., Prasopdee, S., Piratae, S., Khampoosa, P., Suwannatrai, A., Duangprompo, W., et al., 2013. Species-specific primers designed from RAPD products for *Bithynia funiculata*, the first intermediate host of liver fluke, *Opisthorchis viverrini*, in North Thailand. J. Parasitol. 99, 433–437.

Leiper, R.T., 1911. Notes of the occurrence of parasites presumably rare in man. J. London School Trop. Med. 1, 16–19.

Mairiang, E., Elkins, D.B., Mairiang, P., Chaiyakum, J., Chamadol, N., Loapaiboon, V., et al., 1992. Relationship between intensity of *Opisthorchis viverrini* infection and hepatobiliary disease detected by ultrasonography. J. Gastroenterol. Hepatol. 7, 17–21.
Mairiang, E., Laha, T., Bethony, J.M., Thinkhamrop, B., Kaewkes, S., Sithithaworn, P., et al., 2012. Ultrasonography assessment of hepatobiliary abnormalities in 3359 subjects with *Opisthorchis viverrini* infection in endemic areas of Thailand. Parasitol. Int. 61, 208–211.
Matos, M., Alho, A.M., Owen, S.P., Nunes, T., de Carvalho, L.M., 2015. Parasite control practices and public perception of parasitic diseases: a survey of dog and cat owners. Prev. Vet. Med. 122, 174–180.
Murrell, K.D., Pozio, E., 2017. The liver flukes: *Clonorchis sinensis, Opisthorchis* spp, and *Metorchis* spp. In: Rose, J.B., Jiménez-Cisneros, B. (Eds.), Global Water Pathogens Project. Michigan State University, E. Lansing, MI, UNESCO, p. 17. http://www.waterpathogens.org/book/liver-flukes.
Nawa, Y., Doanh, P.N., Thaenkham, U., 2015. Is *Opisthorchis viverrini* an avian liver fluke? J. Helminthol. 89, 255–256.
Ngui, R., Lee, S.C., Yap, N.J., Tan, T.K., Aidil, R.M., Chua, K.H., et al., 2014. Gastrointestinal parasites in rural dogs and cats in Selangor and Pahang states in Peninsular Malaysia. Acta Parasitol. 59, 737–744.
Otake Sato, M., Sato, M., Yoonuan, T., Pongvongsa, T., Sanguankiat, S., Kounnavong, S., et al., 2017. The role of domestic dogs in the transmission of zoonotic helminthes in a rural area of Mekong river basin. Acta Parasitol. 62, 393–400.
Petney, T.N., Andrews, R.H., Saijuntha, W., Wenz-Mücke, A., Sithithaworn, P., 2013. The zoonotic, fish-borne liver flukes *Clonorchis sinensis, Opisthorchis felineus* and *Opisthorchis viverrini*. Int. J. Parasitol. 43, 1031–1046.
Pumidonming, W., Salman, D., Gronsang, D., Abdelbaset, A.E., Sangkaeo, K., Kawazu, S.I., et al., 2017. Prevalence of gastrointestinal helminth parasites of zoonotic significance in dogs and cats in lower Northern Thailand. J. Vet. Med. Sci. 78, 1779–1784.
Ramos, D.G., Santos, A.R., Freitas, L.D., Braga, Í.A., Silva, E.P., Soares, L.M., et al., 2017. Feline platynosomiasis: analysis of the association of infection levels with pathological and biochemical findings. Rev. Bras. Parasitol. Vet. 26, 54–59.
Rojekittikhun, W., Chaisiri, K., Mahittikorn, A., Pubampen, S., Sa-Nguankiat, S., Kusolsuk, T., et al., 2014. Gastrointestinal parasites of dogs and cats in a refuge in Nakhon Nayok, Thailand. Southeast Asian J. Trop. Med. Publ. Health 45, 31–39.
Sadun, E.H., 1955. Studies on *Opisthorchis viverrini* in Thailand. Am. J. Hyg. 62, 81–115.
Salomao, M., Souza-Dantas, L.M., Mendes-de-Almeida, F., Branco, A.S., Bastos, O.P.M., Sterman, F., et al., 2005. Ultrasonography in hepatobiliary evaluation of domestic cats (*Felis catus*, L., 1758) infected by *Platynosomum Looss*, 1907. Intern. J. Appl. Res. Vet. Med. 3, 271–279.
Santos, J.A., Lopes, M.A.F., Schott, A.C., Santos, A.E.S., Porfírio, L.C., Passos, L., 1981. Colangiocarcinomas em gatos com parasitismo de ductos biliares por Platynosomum fastosum. Pesqui. Vet. Bras. 1, 31–36 (English abstract).
Scholz, T., Uhlírová, M., Ditrich, O., 2003. Helminth parasites of cats from the Vientiane province, Laos, as indicators of the occurrence of causative agents of human parasitoses. Parasite 10, 343–350.
Sereerak, P., Upontain, S., Tangkawattana, P., Mallory, F.F., Sripa, B., Tangkawattana, S., 2017. Efficacious and safe dose of praziquantel for the successful treatment of feline reservoir hosts with opisthorchiasis. Parasitol. Int. 66, 448–452.
Sithithaworn, P., Andrews, R.H., Nguyen, V.D., Wongsaroj, T., Sinuon, M., Odermatt, P., et al., 2012. The current status of opisthorchiasis and clonorchiasis in the Mekong Basin. Parasitol. Int. 61, 10–16.

Sriamporn, S., Pisani, P., Pipitgool, V., Suwanrungruang, K., Kamsa-ard, S., Parkin, D.M., 2004. Prevalence of *Opisthorchis viverrini* infection and incidence of cholangiocarcinoma in Khon Kaen, Northeast Thailand. Trop. Med. Int. Health 9, 588–594.

Sripa, B., Brindley, P.J., Mulvenna, J., Laha, T., Smout, M.J., Mairiang, E., et al., 2012. The tumorigenic liver fluke *Opisthorchis viverrini*—multiple pathways to cancer. Trends Parasitol. 28, 395–407.

Sripa, B., Haswell-Elkins, M.R., Sinawat, P., 2003. Histological analysis of gallbladder diseases in relation to opisthorchiasis in endemic areas of Thailand. Acta Trop. 88, 239–246.

Sripa, B., Kaewkes, S., Sithithaworn, P., Mairiang, E., Laha, T., Smout, M., et al., 2007. Liver fluke induces cholangiocarcinoma. PLoS Med. 4, e201.

Sripa, B., Pairojkul, C., 2008. Cholangiocarcinoma: lessons from Thailand. Curr. Opin. Gastroenterol. 24, 349–356.

Sripa, B., Tangkawattana, S., Laha, T., Kaewkes, S., Mallory, F.F., Smith, J.F., et al., 2015. Toward integrated opisthorchiasis control in northeast Thailand: the Lawa project. Acta Trop. 141, 361–367.

Sripa, B., Tangkawattana, S., Sangnikul, T., 2017. The Lawa model: A sustainable, integrated opisthorchiasis control program using the EcoHealth approach in the Lawa Lake region of Thailand. Parasitol Int. 66, 346–354.

Stewart, A.M.E., Gordon, C.H., Wich, S.A., Schoor, P., Meijaard, E., 2008. Fishing in *Macaca fascicularis*: a rarely observed innovative behavior. Int. J. Primatol. 29, 543–548.

Tawfik, A., El-Zamel, T., Herrawy, A., El-Taweel, G., 2015. Fate of parasites and pathogenic bacteria in an anaerobic hybrid reactor followed by downflow hanging sponge system treating domestic wastewater. Environ. Sci. Pollut. Res. Int. 22, 12235–12245.

Taylor, M.A., Coop, R.L., Wall, R.L., 2007. Veterinary Parasitology, third ed. Wiley-Blackwell, Oxford.

Tun, S., Ithoi, I., Mahmud, R., Samsudin, N.I., Kek Heng, C., Ling, L.Y., 2015. Detection of helminth eggs and identification of Hookworm species in stray cats, dogs and soil from Klang Valley, Malaysia. PLoS One 10, e0142231.

Vonghachack, Y., Odermatt, P., Taisayyavong, K., Phounsavath, S., Akkhavong, K., Sayasone, S., 2017. Transmission of *Opisthorchis viverrini, Schistosoma mekongi* and soil-transmitted helminthes on the Mekong Islands, Southern Lao PDR. Infect. Dis. Poverty 6, 131.

Waikagul, J., Thaekham, U., 2014. Approaches to Research on the Systematics of Fish-borne Trematodes. Academic Press, Amsterdam.

Wang, C.R., Qiu, J.H., Zhao, J.P., Xu, L.M., Yu, W.C., Zhu, X.Q., 2006. Prevalence of helminthes in adult dogs in Heilongjiang province, the People's Republic of China. Parasitol. Res. 99, 627–630.

Wang, Y.C., 2012. Examining landscape determinants *of Opisthorchis viverrini* transmission. EcoHealth 9, 328–341.

Wenz-Mucke, A., Sithithaworn, P., Petney, T.N., Taraschewski, H., 2013. Human contact influences the foraging behaviour and parasite community in longtailed macaques. Parasitology 140, 709–718.

WHO, 1995. Control of Foodborne Trematode Infections. Report of a WHO Study Group. World Health Organ. Tech. Rep. Ser. 849, Geneva.

WHO, 2011. Report of the WHO Expert Consultation on Foodborne Trematode Infections and Taeniasis/cysticercosis, Vientiane, Lao People's Democratic Republic 12–16 October 2009, Geneva.

Wongsawad, C., Phalee, A., Noikong, W., Chuboon, S., Nithikathkul, C., 2012. Co-infection with *Opisthorchis viverrini* and *Haplorchis taichui* detected by human fecal examination in Chomtong district, Chiang Mai Province. Thailand. Parasitol. Int. 61, 56–59.

WPRO, 2017. Report of the WHO Expert Consultation on Foodborne Trematode Infections and Taeniasis/cysticercosis, Seoul, Republic of Korea 17–19 May 2017, Philippines.

Wykoff, D.E., Ariyaprakai, K., 1966. *Opisthorchis viverrini* in Thailand—egg production in man and laboratory animals. J. Parasitol. 52, 631.

Wykoff, D.E., Harinasuta, C., Juttijudata, P., Winn, M.M., 1965. *Opisthorchis viverrini* in Thailand—the life cycle and comparison with *O. felineus*. J. Parasitol. 51, 207–214.

CHAPTER FOUR

Social Influence in Liver Fluke Transmission: Application of Social Network Analysis of Food Sharing in Thai Isaan Culture

Waraphon Phimpraphai*[,1], Sirikachorn Tangkawattana[§,||], Suwicha Kasemsuwan* and Banchob Sripa[¶,||]

*Department of Veterinary Public Health, Faculty of Veterinary Medicine, Kasetsart University, Bangkok, Thailand

[§]Department of Veterinary Pathobiology, Faculty of Veterinary Medicine, Khon Kaen University, Khon Kaen, Thailand

[¶]Department of Pathology, Faculty of Medicine, Khon Kaen University, Khon Kaen, Thailand

[||]WHO Collaborating Centre for Research and Control of Opisthorchiasis (Southeast Asian Liver Fluke Disease)/Tropical Disease Research Center (TDRC), Department of Pathology, Faculty of Medicine, Khon Kaen University, Khon Kaen, Thailand

[1]Corresponding author: E-mail: fvetwrp@ku.ac.th

Contents

Advances in Parasitology, Volume 101
ISSN 0065-308X
https://doi.org/10.1016/bs.apar.2018.05.004

Abstract

In northeastern Thai (Isaan) culture traditional raw fish dishes and raw fish-eating habits are common. Eating and sharing meals together among the community's members, especially relatives and neighbours, are a common practice in both daily life and social gathering events. Fish are a significant protein source and are associated with variety of traditional recipes. Cyprinid fish are one of the most preferred fish by Isaan villagers for daily consumption because they are accessible and affordable. Consumption of these fish probably causes the persistence of high endemicity of human liver fluke infection, particularly with *Opisthorchis viverrini*, in northeast Thailand. Because the consumption of raw cyprinid fish is a well-documented risk factor for liver fluke infection, sharing of risky raw fish dishes may influence disease transmission through a community. Social network analysis was used to investigate fish and fish-based meal sharing among household members in Isaan villages in liver fluke endemic areas. The findings from three studies confirmed the persistence of traditional Isaan raw fish consumption and food-sharing practice. Social connections via food sharing among villagers played an important role in liver fluke infection and transmission dynamics as a risk factor. Thus these sociocultural factors should be taken into account in designing strategies for control of opisthorchiasis and other food-borne illnesses at the community level.

1. INTRODUCTION

Human liver flukes are food-borne trematodes that remain a significant public health problem in various parts of Southeast Asia, especially in the Lower Mekong Basin (Sripa et al., 2010; Sithithaworn et al., 2012). Liver fluke infection is acquired by eating raw fish, a behaviour related to traditional fish recipes. In northeast Thailand (Isaan) where *Opisthorchis viverrini* is endemic, intensive and continuous liver fluke control programs have been carried out as public health service activities. These strategies have coincided with lifestyle transitions that changes the communities from traditional agricultural life to modern urban industrial life, which potentially resulted in a decrease of liver fluke prevalence (Jongsuksuntigul and Imsomboon, 2003). However, prevalence data still indicate high *O. viverrini* infection rates of up to 50% in certain Isaan provinces, where infection has tended to persist in highly endemic areas (Sithithaworn et al., 2012).

This may be caused by failure to inspire communities to long-term change, or it may be due to the fact that control strategies did not take into account the various effects of social and ecological factors on OV transmission dynamics. Effective and sustainable *O. viverrini* control needs integrative, multisectoral and multidisciplinary approaches to overcome these obstacles (Sripa et al., 2015). Focussing on social determinants, including local culture, beliefs and behaviour, the *Lawa model* indicated that local culture can determine behaviours associated with fishing, food preparation and the consumption of undercooked fish. All these practices are deeply rooted as part of the native rice-fish culture in the region (Sripa et al., 2015). Social network analysis (SNA) is an approach based on the relationships between *nodes* and *ties* which describe interactions between individuals within a group, improving our understanding of the behaviour of a group (Martínez-López et al., 2009). Networks diagrams are regularly represented as nodes and links between these nodes. This technique can be used for mapping the distribution of freshwater fish within a community and characterizing the pattern of food sharing, particularly of traditional fish dishes, among villagers in an endemic area of human opisthorchiasis.

2. SOCIOCULTURAL FACTORS RELATED TO HEALTH

Perception of the world by particular people is deeply influenced by sociocultural factors, including the beliefs, customs and practices within cultures and societies that affect the thoughts, feelings and behaviours of its citizens (http://psychology.wikia.com/wiki/Sociocultural_factors). These sociocultural aspects include religious beliefs, culture, attitudes, role of the family, societal status and perception on various issues. Our understanding of these factors can be applied in different fields, such as psychology, design, marketing, business and the health professions.

Health is influenced by various factors, including genetics, human behaviour, the external environment, sociocultural factors and healthcare access. Sociocultural variables affect health at critical stages throughout the course of life, and there are multiple levels of exposure. These include socioeconomic status (SES), ethnicity, gender, poverty, social network, social support, the psychosocial work and social environments (Hernandez and Blazer, 2006).

2.1 Effect of Socioeconomic Status on Health

The association between health and SES is extensive in all societies which results in diverse health outcomes (Case et al., 2002). Educational attainment, income and occupational status have been used as SES indicators, separately or combined. Education has been linked with health outcomes through disease morbidity, mortality, health behaviour and functional limitations, which does not necessarily imply causation (Hernandez and Blazer, 2006). Knowledge and skills acquisition have been hypothesized as causal pathways for better health outcomes, including health promotion, health literacy improvement, higher social status and prestige and employment prospects (Cutler and Lleras-Muney, 2006). The effect of income on health is complicated and difficult to measure because of its dynamic and mutual relationship. The causal pathway of income and health linkage suggests that health is strongly associated with level of household income, with lower income relating to worse health status (Case et al., 2002). On the other hand, there is evidence that health status is improved in higher income groups (Costello et al., 2003). Income certainly allows individuals to pay for necessary goods and services, allowing them to maintain their health status, and provides a psychological sense of control and mastery over their environment. Thus higher income is also related with healthier behaviour (Case and Paxson, 2002). Occupational status is the standard component of SES, which includes different levels of prestige, authority, power and resources. Illness is a crucial cause of downward occupational mobility. This means an association between occupational status and health can partly reflect reverse causation (Hernandez and Blazer, 2006). Education, income and occupation can mutually influence and interplay with each other throughout the life course of individuals, determining health outcomes at multiple levels of social organization (Hernandez and Blazer, 2006).

2.2 Social Network, Social Support and Health

Social determinants of health are important in terms of the scope, power and quality of social connections. Steady linkages are necessary for both material resources, such as food and warmth, and nonmaterial resources, including love and security, that are significant for human development (Berkman et al., 2000). Social relationships are characterized by two particular variables: social network and social support. A social network is defined as a person-centred social ties Web that assesses social linkage structure including size, density, boundedness and homogeneity (Berkman et al., 2000). Social support refers to different types of assistance that people receive from social networks and can be identified as instrumental, emotional and informational support (Berkman et al., 2000). Social network or social support is bidirectional. These are associated with health that can have a positive or negative influence concurrently. However, temperament and personality may be confounders that affect these relationships. The effects of social network and social support on health can be extended to the community level. The resources and other social contexts that are available for community members have been defined as social capital according to the community-level mechanisms such as informal social control, collective efficacy, collective socialization and social contagion (Kawachi et al., 2004). Therefore social capital and social cohesion are significant characteristics of the sociocultural environment that affects health outcomes in different ways from disease onset to progression and survival.

3. SOCIOCULTURAL INFLUENCES ON FOOD AND EATING BEHAVIOUR

3.1 Food as Social Function, Food Production and Distribution

Food is a crucial society element and plays important roles not only as a source of nutrition but also in daily life, beliefs and socioeconomics. Interpersonal relationships between community members and their sociocultural environments can be expressed and established by food (Ma, 2015). Various messages can be transmitted through different kinds of foods, indicating the relationship proximity among people (Fieldhouse, 1995). SES can be indicated by the consumption of rare and expensive food, as well as group characteristics indicated by families, races, religions, regions or countries (Ma, 2015). People celebrate important events or festivals by eating special food. However, different societies and cultures potentially influence particular food traditions (Fieldhouse, 1995).

Food cultivation, harvesting, production, serving, consumption and eating habits are culturally defined. Women in most societies play an important role in food production, selection, purchase and processing (Ma, 2015). Women usually act as cooks and food providers for their families, so engagement in food market trade and decision-making relating to the type, quantity and quality of available food are often their responsibility. Women are also responsible for food distribution within the family according to two modes: demand and contribution. The demand mode has been defined by food distribution based on the distinct physical demand for nutritional intake among family members. The contribution mode is indicated by the level of contribution to the family by each family member. Priority in food selection and having the largest amount and best part of food are reserved for the members who earn more money. Gender and age also play different roles in food distribution and have an effect on the health of family members (Ma, 2015).

3.2 Food and Culture

Eating and sharing meals among the community's members are basic elements of a society, and no culture has been established without food sharing. Daily routine and ritual structure at both the formal and informal life stages is associated with food transfer. Food also creates a common vision for culinary culture sharing. The fundamental rule in food culture is identifying which food is edible. Biological and geographical conditions, as well as cultural norms, define the eating behaviour within a particular community. However, ideas and traditions regarding edible foodstuff differ between nations or social classes (Nordström et al., 2013). Food leads to mutual obligations among relatives and the community. Relationships are usually started by an offer of food sharing, with refusal of meal sharing resulting in the rejection of a social connection. Food presentation is relevant to the level of power of the sharer and is also associated with the conflicts between social classes, ethnic groups and nations (Nordström et al., 2013). Food can be used to show respect to an individual's social rank. It is a crucial part of the links between genders because, traditionally, females act as family food providers. Women may be excluded from public power but nevertheless imprint power, values and morals because of their role of cooking for their family.

3.3 Eating Behaviour and Social Influences

Sharing a meal with the family, friends and colleagues is a common activity. Through various means, other social group members affect particular persons through their food intake and selection (Oh et al., 2014; Xayaseng et al., 2013). Social context potentially influences eating behaviour because different people eat differently when they are with other people compared with when they are alone. In such situations food selection has the tendency to harmonize with close social connections in society because of behavioural adaptation and rewarding. Thus other people's behaviour, cultural expectations, sharing and environment set up suitable eating norms. Eating norms are certain codes of conduct or guidance for appropriate eating, including both food choice and the amount of food intake (Higgs, 2015). Based on social comparison, people who are perceived to be relevant to particular social group are more likely to follow an eating norm. During the process of eating, action synchronization, consumption monitoring and modified food preferences are involved with norm matching (Higgs and Thomas, 2016). Different groups of people can be characterized by their eating patterns. These are embedded in their social relations and reflected in the fundamental political, economic and meaning system of their places (Delormier et al., 2009).

4. FOODS AND FOOD-CONSUMING BEHAVIOUR IN THAI ISAAN CULTURE

The Mekong River passes through six countries: China, Myanmar, the Lao PDR, Thailand, Cambodia and Vietnam. Thirty-seven percent of the land area of Thailand lies within the Lower Mekong Basin. The major subbasins in northeast Thailand are the Kok River Basin, the Chi River Basin, the Mun River Basin and the Ang Sakon Nakhon Basin (Na Mahasarakarm, 2007). Despite extensive dam building for water storage and irrigation purpose, much of this region is flooded annually, either from rainfall or floodwaters. Dams, water reservoirs and seasonal wetlands, especially rice fields, are created in vast areas which are the basis of much fishery production. Fisheries are important particularly to the rural people who live in the basin and rely on fish as a crucial part of their daily diet (Na Mahasarakarm, 2007).

Isaan is located on the Khorat Plateau, bordered by the Mekong River to the north and east and by Cambodia to the south. To the west it is separated from Northern and Central Thailand by the Phetchabun mountain range. In Isaan rice paddy fields comprise 95% of the land classified as 'wetlands'. The other significant wetland habitats are lakes and ponds (natural or artificial) and swamps (including backswamps, grasslands and marshes) (Na Mahasarakarm, 2007). Swamps are known for their high fish productivity. Fish species composition varies between water bodies, but the major species belong to the carp family, Cyprinidae, and include *Cirrhinus jullieni*, *Barbodes gonionotus*, *Puntioplites proctozysron*, *Micronema* sp., *Channa striata*, *Hemibagrus nemurus*, *Oxyeleotris marmorata* and *Clupeichthys aesarnensis* (Sayasone et al., 2007; Vonghachack et al., 2017). Besides the various fish species, other aquatic animals such as frogs, small shrimps, crabs, tadpoles, clams and snails also make an important contribution to the wild-caught freshwater diet of the Isaan population (Na Mahasarakarm, 2007). This is consistent with the Lawa Lake study in the Chi River Basin which showed that three species of Cyprinidae are the most abundant fish caught and consumed (Kim et al., 2017). Cyprinid fishes are mostly preferred by

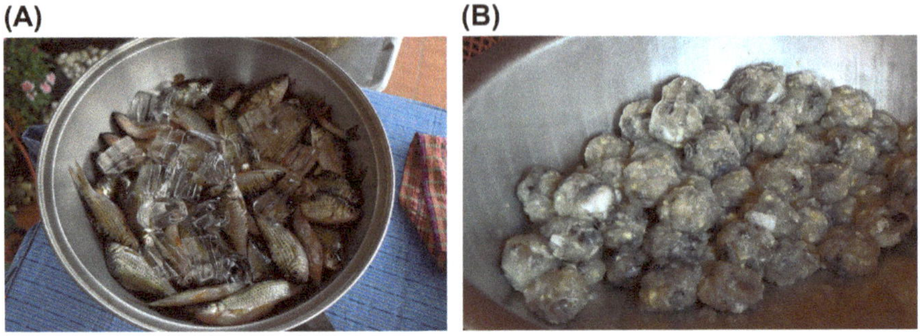

Figure 1 Cyprinid fish (A) and *Pla-som* (B) from Lawa Lake area, Khon Kaen Province.

Isaan villagers for daily consumption because they are easy to catch or locally sold at an affordable price. If caught, they can then be distributed for free to family and neighbours (Kim et al., 2017).

Fish and fermented fish products (Fig. 1) are crucial for Southeast Asian diets as the major protein source in most meals (Mutsuyama, 2003). Many rural Southeast Asians raise fish in ponds near their homes and rice fields. Since ancient times, rice and fish have been the staple diet for the people in this region. Fishing is commonly practiced in most rural communities, particularly those located near water bodies. Fishing skills are traditionally handed down from parents to children, generation after generation, so that this local knowledge and expertise remain alive. Fishing was formerly restricted to providing food for fishermen and their family members. Any excess amount was turned into processed products such as dried or fermented fish for household consumption and sharing with relatives or friends. In addition, fish and fish products were sometimes used as commodities to barter for other goods (Mutsuyama, 2003).

Isaan foods and eating practices have shifted from traditional to more modern lifestyles. However, traditional consumption of raw fish continues despite the improvements in eating habits and health education programs (Grundy-Warr et al., 2012). In addition, there remains persistence of incorrect beliefs that adding lime and/or drinking alcohol when eating raw fish will reduce risk of food-borne parasitic infection (Kim et al., 2017; Xayaseng et al., 2013). Isaan iconic raw fish dishes are the expressions of their culture: *Koi-pla* is a spicy minced fish salad, eaten soon after preparation; *Pla-som* is moderately fermented and can be stored for a few days to weeks and *Pla-ra* is an extensively fermented, highly salted fish dish that can be stored for at least 2–3 months (Kaewpitoon et al., 2008). Traditional Isaan dishes prepared from freshwater cyprinoid fish, such as *Koi-pla* and *Pla-som*, are the usual sources of liver fluke infection in northeastern Thailand and Lao PDR (Sripa et al., 2011; Sithithaworn et al., 2012). One of the difficulties relating to disease prevention and control is to dissuade risk populations from eating raw or fermented fish and encourage them to properly cook this food to kill the parasites (Sithithaworn and Haswell-Elkins, 2003; Andrews et al., 2008; Sripa and Pairojkul, 2014; Shin et al., 2010; Sriraj et al., 2016). Today the consumption of raw fish dishes is discouraged as villagers and public health officials consider them unhealthy (Sithithaworn et al., 2012; Kim et al., 2017).

5. LIVER FLUKE INFECTION IS A FOOD-BORNE DISEASE

Liver flukes causing fish-borne zoonotic trematodiasis cause major public health problems worldwide. The number of people currently infected with liver flukes (family Opisthorchiidae) exceeds 18 million (Chai et al., 2009), but it is estimated that more than half a billion are at risk of infection (Lima dos Santos and Howgate, 2011). Humans contract liver fluke infection by the consumption of raw or undercooked infected fish (World Health Organization, 1995). Lima dos Santos and Howgate (2011) indicated that these food-borne parasitic zoonoses have increased significantly for various reasons: (1) the development of new and improved diagnosis, (2) increased raw fish consumption in traditional dishes, (3) increased consumption of regional fish dishes such as sushi, sashimi, ceviche and carpaccio based on raw or minimally processed fish, (4) the growth of international trade in fish and fish products and (5) the striking development of aquaculture, particularly in Asia (SEAFDEC, 2017).

O. viverrini is endemic in Thailand, Lao PDR, Cambodia and the southern parts of Vietnam, with an estimated 10 million plus people in this region being at risk of infection (World Health Organization, 1995; Sripa et al., 2010; Sithithaworn et al., 2012). *O. viverrini*, which is of great medical importance, is highly endemic in Thailand. Infection with this liver fluke is particularly widespread in the north and northeastern regions of this country (Sripa et al., 2003; Kaewpitoon et al., 2008), where an estimated six million people are infected (Jongsuksuntigul and Imsomboon, 2003; Sripa et al., 2003). One reason for the persistent high infection rates in this region is likely its cultural and ecological identity, where rice cultivation and a strongly embedded raw fish consumption culture are present (Grundy-Warr et al., 2012) and create ideal conditions for sustained transmission (Sripa et al., 2007; Sripa et al., 2011, 2015). The transmission of liver fluke is associated with behavioural patterns that are determined by sociocultural and economic conditions. Raw fish consumption usually occurs among inhabitants living around lakes, streams and ponds (Hung et al., 2013). Food safety education may be the most appropriate strategy to control the spread of the disease due to its inherent link to human behaviour (Ziegler et al., 2011). However, changing eating habits is challenging because these are deeply rooted in cultures.

6. SOCIAL NETWORK ANALYSIS

6.1 History of Social Network Analysis

Social network was mentioned in 1932 to describe an epidemic of runaways at the Hudson School for Girls (Moreno, 1934). The reason for the increasing number of runaways was more dependent on the social position of the individuals affected in their social network than on their individual personalities and motivations. Sociometric techniques were used to elicit and graphically present the individual's feelings such as 'like' or 'dislike' of the other individuals. In the 1940s and 1950s the concept of social networks was developed in two ways. First, matrix algebra and graph theory were used for discovering emergent groups in network data to formalize basic social—psychological concepts. Second, laboratory experimentation of networks programs was developed (Borgatti et al., 2009). The effects of different communication network structures on the speed and accuracy of problem-solving were studied by researchers from Massachusetts Institute of Technology (MIT) (Borgatti et al., 2009). They found that a more centralized network structure outperformed decentralized structures (Leavitt, 1951). In human networks, information tends to be conveyed from peripheral members to the most central nodes, so those systems with the shortest distance between all nodes from the distinct integrator nodes performed fastest (Borgatti et al., 2009). In 1978 Pool and Kochen published the 'small world' concept that determined the chance that randomly selected pairs of people will know each other. Based on mathematical models half of the total number of pairs among USA population could be linked with not more than two mediators (Pool and Kochen, 1978). Concordance with this finding, the famous concept called *six degrees of separation* was developed by Milgram (1967). Network analysis was also used by sociologists to study the concrete relations between people, as well as to represent community structure (Borgatti et al., 2009). Community network structure analysis has become an increasingly significant research tool in social sciences (Wellman et al., 1996). Social network application has been successfully used in anthropology to study patterns or networks of relationships, kinship and for the network-based explanation of a wide range of outcomes (Borgatti et al., 2009). According to strong or weak ties theory close contacts between persons form cluster(s) of strong ties which subsequently leads to the redundancy of information sharing (Granovetter, 1973). Social capital theory was derived from this idea to determine whether these connections enable people to access resources leading them to a better situation. In the 1990s network analysis research increased in a number of fields, including physics and biology, and also to various applied fields such as management consulting, public health and crime/war fighting (Borgatti et al., 2009).

6.2 Social Network Terminology and Description

A network is a collection of units of interest which are generally called *nodes*, *vertices* and *actors* in physics, mathematics and social sciences, respectively. *Nodes* are members of the network representing distinct individuals or collective units. Attributes are the nodes' properties, qualities or characteristics that belong to them as an individual or group. Nodes are connected to each other through different types of relationship (Dubé et al., 2011). These links also called *ties*, *edges* or *arcs* and can be classified into four basics types of dyadic relations: similarities, social relation, interactions and flows (Borgatti et al., 2009). These ties can be informal or formal, and multiplexity occurs when actors have multiple ties with other nodes. *Edges* refer to reciprocal or undirected links, whereas *arcs* refer to the links that are unidirectional or directed (Wasserman and Faust, 1994). *Edges* and *arcs* describe the network representation as undirected or directed networks and are defined by binary or other values based on their relationships as characterized by the study considered. The total number of *nodes* determines the size of a network. SNA can be illustrated either in matrix format or as graphs. However, using the matrix format allows the execution and calculation of a range of measures that provide network descriptive statistics (Dubé et al., 2011).

Networks can be classified in four main types (Hawe et al., 2004) including the following:

6.2.1 *One-mode networks*: the connections among a single set of similar nodes, such as exchange of information among health volunteers in the village.

6.2.2 *Two-mode networks*: the connections among two different sets of nodes used to explore the relationship between nodes and a series of events. Ties exist distinctively only between nodes that are considered more responsible for tie creation than others such as the links between researchers and disciplines or specialities related to their field of research.

6.2.3 *Sociocentric or complete networks* are composed of the connected ties among single, bounded community members such as a network of medical doctors in a hospital.

6.2.4 *Egocentric or personal networks* are mentioned based on the focal node's perspective only. The ties directly connect the focal node (ego) with others (ego's alters) and also ego's views on the ties, if any, among the alters. For example, in the ego network of health-related information sharing among village members, the ego would be the health volunteers who are asked if the information is directly shared with other villagers. These alters are subsequently asked to identify the next nodes with whom the information is shared.

6.3 Measure of Centrality and Cohesiveness

Individual-level network data are collected, but these data can also be analyzed at the structural level. In general, network data are typically entered into a database by node similarity or distance matrix. Existence of ties is defined by '1', whereas no tie is defined by '0'. The ties can be labelled as valued data if ties' intensity is examined. The strong and weak ties represent the large number of connected nodes in a similarity matrix and the larger number of distances between nodes in a distance matrix, respectively. These data can be illustrated in graphs visualized and analyzed using a variety of special network analysis software packages (Hawe et al., 2004) , such as UCINET 6 (Analytic Technologies, Harvard), Pajek (http://vlado.fmf.uni-lj.si/pub/networks/pajek/default.htm) and StOCNET (version 1.4, ProGAMMA/ICS, Groningen).

6.3.1 Measure of Node-Level Centrality

To conduct SNA, one of the main objectives is to identify the important nodes or central nodes in the network. The nodes are then characterized using three measures: degree, betweenness and farness (Wasserman and Faust, 1994). The node degree shows the number of contacts per node, as well as the number of linked nodes in an undirected network. The degree value can be categorized into out-degree, which is the number of ties that originate from each node, and in-degree, which is the number of ties that each node receives in a directed network (Wasserman and Faust, 1994). The node betweenness reveals the frequency of the shortest 'path' between pairs of nodes in the network, indicating their important role for linking a large number of pairs and fragmentation of the network. The node farness demonstrates topologically the distance from particular nodes to all other nodes in the network; it is the inverse of node closeness (Dubé et al., 2011).

6.3.2 Measure of Network-Level Cohesiveness

Network-level cohesiveness estimates the coherence level of the whole network. The measured parameters consist of network density, fragmentation, average path length and clustering coefficient (CC) (Dubé et al., 2011). The network density refers to the proportion of actual links out of all possible links, and the value can range from 0 to 1. Network fragmentation demonstrates the proportion of unreachable pairs or the pairs that the path does not link (ranges from 0 to 1). The network average path length represents the average shortest distance or number of steps among reachable pairs in the network. Social scientists apply this measure to explain the connected world leading to the *six degrees of separation* concept (Milgram, 1967). The network CC is a principal measure because of its application to network characterization such as small world. The CC is defined as the proportion of one's neighbours in the network who are also neighbours of one another (Watts and Strogatz, 1998). Characteristics of small-world networks include high CCs and short average path length with few numbers of topologically long-distant cluster links in the network (Dubé et al., 2011).

6.4 Social Network Analysis Reflects Disease Transmission Patterns

Social networks have been used as a map of direct contact that facilitates infectious disease transmission. Focussing on the person-to-person transmission perspective, individual-level risk is associated with relationships between society members (Newman, 2002). Social cohesion and organization resided in the networks influence on community-level risk of infection with infectious diseases (Pahl-Wostl et al., 2008). Social connectedness has long been considered as contact in disease transmission models (Klovdahl et al., 2001; Meyers et al., 2003; Bansal et al., 2010) and as a protective factor for chronic disease (House et al., 1988; Berkman and Glass, 2000; Cohen et al., 2007). The mechanisms of social organization and action for enhancement or inhibition of pathogen transmission via the environment are necessary for understanding the role of the social network in infection risk (Zelner et al., 2012).

The crucial role of a complex network of transmission routes has been revealed in a variety of infectious disease outbreaks. Epidemiologists have studied the roles of social networks in disease spread over the last two decades (Riolo et al., 2001; Hawe et al., 2004; Trostle et al., 2008). SNA was applied for a number of epidemiological studies focused on risk potential networks and their social relationship to sexually transmitted infections (Doherthy et al., 2005) or for the transmission of human immunodeficiency virus (HIV) through a range of contacts in a network, including sexual contact and intravenous drug use infection (Klovdahl et al., 2001; Friedman et al., 2000; Rothenberg et al., 2001; Riolo et al., 2001). The contact network for the severe acute respiratory syndrome (SARS) outbreak has been mentioned because of the long distance travel that caused its local and later worldwide spread (Peiris et al., 2003). The foot-and-mouth disease outbreak in the UK, similar to SARS outbreak, was influenced by long-range movements, in this case of cattle, both nation-wide and then locally (Kao, 2002; Mansley et al., 2003). SNA can be used to conceptualize and demonstrate complex networks for these infections (Christley et al., 2005). Centrality measures have been used to determine the importance of each individual within a population according to the network criteria. These criteria include the number of contacts made by each individual, the number of steps from each individual to all others and the frequency that an individual presents a pathway among others (Wasserman and Faust, 1994). Centrality measures have already been applied to study HIV and syphilis infection in humans (Bell et al., 1999; Rothenberg et al., 2001) and *Mycobacterium bovis* infection in captive possums (Corner et al., 2003) to indicate the individuals who could be associated to the disease transmission. There are many network parameters associated with infectious disease risk. Network parameter specification will help identifying intervention targets, such as control measures and surveillance, and in developing mathematical models and simulations to predict disease transmission through the networks (Bell et al., 1999; Lloyd and May, 2001; Liljeros et al., 2001). Networks involving person-to-person contact could also demonstrate the enhanced transmission of several diseases, including influenza (Sattenspiel et al., 2000), and *Staphylococcus aureus* infection (Lowy and Miller, 2002). Moreover, SNA has also been used in the study of the distribution of obesity (Christakis and Fowler, 2007) and diarrhoeal disease transmission via food sharing (Bates et al., 2007; Trostle et al., 2008; Zelner et al., 2012).

7. APPLICATION OF SNA TO FOOD-SHARING BEHAVIOUR ASSOCIATED WITH LIVER FLUKE INFECTION IN THAI ISAAN CULTURE

Three community-based studies were conducted in Isaan region at the centre of Lower Mekong Basin to elucidate the association of raw fish meal practice and food-sharing habit with risk of *O. viverrini* infection and transmission. Stakeholder analysis was considered to confirm the key informants for community engagement and information gathering including the local authority officers, village headmen, health volunteers, fishermen, fishmongers and villagers. Multimethod approach that combined both qualitative and quantitative methods was used to collect all relevant information. Qualitative methods such as focus group discussion, in-depth interview and participatory epidemiology were used to investigate sociocultural basis on fish recipes, raw fish consumption and food-sharing behaviour of Isaan villagers. Quantitative methods including questionnaire survey, geographic information system (GIS) and SNA were used to characterize network that shares fish-based dishes as well as geographical distribution of freshwater fish and estimate risk of liver fluke infection associated with raw fish meals sharing among household members under village level.

7.1 Food-Consuming Behaviour in Endemic Area

Food-eating and food-sharing behaviours were determined among villagers living in an *O. viverrini* endemic area around Lawa Lake, Khon Kaen Province (Kim et al., 2017; Phimpraphai et al., 2017) and in Maha Sarakham Province (Saenna et al., 2017) where many kinds of cyprinoid fish are abundant. Fish, especially cyprinids, were the most important protein source and contributed significantly to the daily diet because of their palatability, availability and affordability. Most of the local villagers prepared food by themselves, and the favourite fish dishes for all household members were prepared by deep-frying, boiling and using fish-based chili sauces. The three most popular traditional raw fish dishes of Isaan cuisine are *Koi-pla*, *Pla-som* and *Pla-ra*. Gender is one of the important factors for *O. viverrini* infection, and the prevalence in males is higher than in females (Sriamporn et al., 2004). But, the effect of gender on raw fish consumption was masked by using the household as a unit of interest. The information from participatory epidemiology revealed that males had more opportunities for eating raw fish with drinking alcohol during social gatherings, but both genders ate raw fish dishes such as *Pla-som* on a daily basis, which is similar to the finding of Grundy-Warr et al. (2012). Because many public health activities and campaigns to prevent consumption of raw fish have been introduced aimed at preventing *O. viverrini* infection, the villagers are likely to be aware of the risk of eating raw or undercooked fish dishes such as *Pla-som* and *Koi-pla*. Nevertheless, eating raw fish still remains common in particular groups of people because of the long-standing tradition of preparing raw fish dishes and the consumption culture in Isaan. In addition, spicy papaya salad (*Som-tam*) prepared with *Pla-ra* is still consumed regularly among people in the endemic areas.

Additional information was described by the participants from Lawa Lake study about the transition of traditional Isaan diet to be of more western style which had happened since last 10—15 years. Fast food or ready-to-eat meals such as instant noodles, bread, coffee and hotdog have easy access because of the better transportation, incomes and village-based convenience shops (Kim et al., 2017). These will have an impact on food consumption practices in Isaan region that is becoming modernized compared with the past year.

7.2 SNA on Fish and Food Sharing and Liver Fluke Infection in Lawa Lake, Khon Kaen and in Maha Sarakham

Wetland livelihoods of the local communities are associated with multiple and complex sociocultural and ecological factors related to liver fluke transmission and persistence (Friend, 2007). Wetland ecosystem produces and provides water, food and traditional medicines that are crucial for local livelihood practices. Various kinds of fish can be either caught naturally or cultivated in this rich ecosystem. Cyprinidae was the family of fishes found most abundantly which had been caught naturally. The preference for raw cyprinid fish consumption in Isaan and Lao PDR is underlined by these factors (Grundy-Warr et al., 2012; Xayaseng et al., 2013). In Khon Kaen Province *O. viverrini* prevalence levels range from 2% to 71% (Sripa et al., 2015).

Food sharing, especially of cyprinid fish and raw fish dishes, happens regularly during daily life, as well as at social events, among family members, relatives and neighbours. Grundy-Warr et al.(2012). Focussing on the application of SNA to our understanding of both fish distribution and fish-based meal sharing (Fig. 2) with respect to liver fluke infection in Thailand, two studies were carried out at the Lawa Lake region in Khon Kaen Province (Kim et al., 2017; Phimpraphai et al., 2017). Lawa Lake is a semiartificial reservoir and wetland ecosystem. It serves as an ideal habitat for *Bithynia*

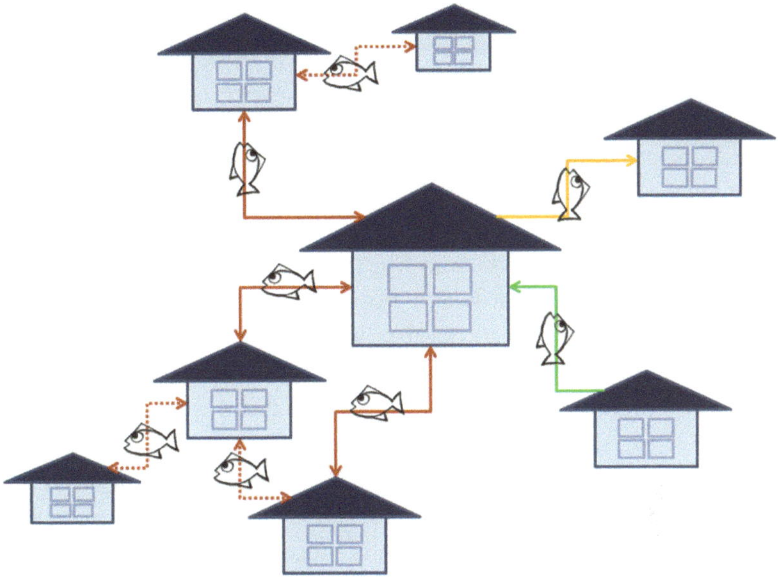

Figure 2 Social networks of freshwater fish sharing and fish-based dishes sharing among village households. Direction of food sharing indicated by *arrow* and *colours*, red: both sharing-in and sharing-out; yellow: sharing-out only; green: sharing-in only.

snails and cyprinid fish, the first and second intermediate hosts of *O. viverrini*, respectively. The area is known to be a highly endemic area for opisthorchiasis in humans (Sithithaworn and Haswell-Elkins, 2003; Sripa et al., 2011). Another study was carried out in Maha Sarakham Province located at the centre of the Lower Mekong Basin where the gathering, preparing and sharing practices of fish dishes are deeply rooted in the local culture (Saenna et al., 2017).

A multimethod study was carried out by Kim et al. in 2017 to explore the sociocultural determinants of raw fish consumption behaviour among Isaan people at Lawa Lake region. Distribution routes for potentially parasitized fish were studied using partial network analysis. In-depth interviews using SNA questionnaires, which highlighted in direction of catching and selling fish, were conducted with 46 fishing-related participants including fishermen, fish sellers/buyers, middlemen, *Pla-som* makers and fishmongers to define the fish distribution networks and movement of local fish from this *O. viverrini* endemic lake to every actor among market chains. The results revealed the fish distribution and logistics network and elucidated the key roles of middlemen, *Pla-som* makers and local vendors in terms of the important links for fish distribution between fishermen and consumers. According to fishermen's opinion, larger fish are more economically attractive than smaller fish, so these kinds of fish are mostly transported away from local communities to town markets. Smaller fish, particularly cyprinid fish, remain in the local village for sharing and selling at a low price to fishermen's families and friends, which maintains the continuity of *O. viverrini* infection in these villages (Kim et al., 2017).

A second study was carried out in 2017 by Saenna et al. in Maha Sarakham Province, which is located 35 km southeast to Khon Kaen municipality. The area is a typical lowland floodplain along the Chi River. Ego network analysis was used to determine raw fish sharing in 2 of 12 rural villages. The selection criteria for these two villages are they have high *O. viverrini* prevalence (34.6% and 48%) and distinct spatial layout of the households. One village has a closely clustered pattern of households, but the other has a linearly pattern such that households disperse along the road. The researchers expected that different spatial arrangement of each particular village might result in dissimilar social interaction and food-sharing patterns. Questionnaires consisted of two phases; the first phase was designed for *O. viverrini* community-level intervention, and the second phase had a supplementary SNA questionnaire that exclusively focused on questions about fish dish sharing. Partial network analysis was carried out using nodes defined as households and ties defined as food-sharing relationships based on four types of raw fish dishes *Koi-pla*, *Pla-som*, *Pla-ra* and *Pla-jom*. Household locations were determined by a handheld global positioning system receiver (Garmin Ltd.; Olathe, KS, USA), and household maps were created using Quantum GIS, version 1.8.0 (http://www.qgis.org/en/site/). This information was then used to visually characterize the spatial pattern of households in each village. Statistical analysis of raw fish sharing and *O. viverrini* infection risk associated with exposure to raw fish dishes were performed using R statistical package, version 3.0.3 (R Core Team, 2013). Raw fish-sharing networks were constructed, and the degree of food sharing (DFS), which is defined as the number of the household members who share food together regardless of providing or receiving food, was measured using the igraph R package. These networks were complex and showed an apparent association between household food sharing and *O. viverrini* infection prevalence. Information from the network analysis demonstrated that a household's members who were positive for *O. viverrini* infection also tended to consume a greater variety of raw fish dishes. The direction of the ties showed that providers were household members who fished and prepared fish dishes, whereas the recipients were household members who did not fish, mainly elderly or disabled individuals. A statistically significant association was showed between DFS and *O. viverrini* infection and other risk factors such as raw fish consumption and types of raw fish dishes. However, the result revealed the different networks that share raw fish dishes which illustrate more complexity or greater number of sharing in a clustered pattern village than in a linear pattern village. In addition, social determinants of raw fish dish sharing were observed. Three factors associated with these behaviours were expressed: enjoyment of raw fish dishes among relatives and friends, alcohol drinking and access to the fish used in raw fish dishes (Saenna et al., 2017).

A third study was conducted as a part of the Lawa Model established by the Tropical Disease Research Center of Khon Kaen University (Phimpraphai et al., 2017). The model is an integrated liver fluke control program using a transdisciplinary approach to address interventions for *O. viverrini* infection in the Lawa wetland (Sripa et al., 2015). Participatory epidemiology was used to explore food sources and food-eating habits among key informants, consisting of village leaders, fishermen, health volunteers and villagers. Related information, including livelihood, agricultural patterns, protein sources and fish recipes, on daily consumption and social gatherings was also collected (Fig. 3). Two villages with the lowest and highest *O. viverrini* infection (29.17% and 88.17%, respectively) at the household level were recruited to investigate food sharing among households. Ego network was applied in these two selected villages when food sharing occurred directly between the focal households (ego) and other households (alter) as well as the ego's perspective on food sharing among alters and other nodes. The ego nodes were defined as the households randomly selected from 10% of total households in each village including equal number of either positive or negative *O. viverrini* infection households. Alter nodes were all possible households who share food with egos and also with others. The links between egos and alters were indicated by their food-sharing or trading activities. The link direction was considered by the direction of sharing-in (in-degree) and sharing-out (out-degree). Questionnaire surveys including social network and attributes data, such as *O. viverrini* infection, fish sharing, food sharing, freshwater fish recipes and raw fish-eating behaviour, were conducted with all egos and their alters. The data included food-sharing events in the previous year, but data on raw fish-eating behaviour were from the previous 10 years as well as the previous year. The DFS and trading among households in each village's network was assessed. UCINET, version 6.530 (Borgatti et al., 2002)

Figure 3 Ranking of favourite traditional Isaan fish dishes for daily consumption. According to daily fish-based meals, *Pla-ra* is the most favourite fish dish, whereas the second to fifth ranks are given to cooked fish dishes including sundried and deep-fried fish, steamed fish, Isaan style fish soup and grilled fish, respectively. Traditional raw fish dish such as *Koi-pla*, *Pla-som* and *Pla-jom* were still preferred in small number of villagers.

was used to generate network diagrams and calculate network parameters. Different degree numbers between networks were also adjusted by normalized data.

In each village the results revealed that most of the inhabitants were rice farmers and freelance workers. Lawa Lake, where cyprinoid fish are abundant throughout the year, was their main food source, particularly during rainy season (September to November). Fish and eggs were the most important protein sources of the villagers because of their palatability, availability and affordability. Raw fish dishes such as *Pla-som* and *Koi-pla* were avoided, but spicy papaya salad with *Pla-ra* was still regularly consumed. Food sharing usually happened during particular social events and more regularly among peers and relatives (Phimpraphai et al., 2017).

Ego network analysis was conducted to explain all possible links of food sharing among the households between these two villages with different infection rates. To formulate an ego network, 179 out of 353 households were enrolled from these two villages. A graphical illustration of ego networks of food sharing between the two villages is shown in Fig. 4. A sparse pattern of small group sharing is found in the high infection village, whereas a larger and more connective pattern was present in the low infection village. Between these two villages, food sharing was similar in numbers

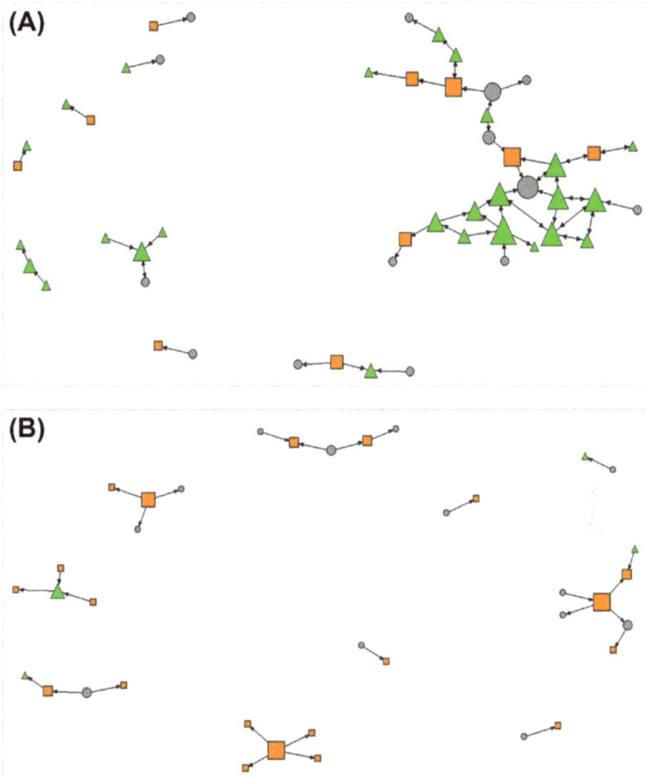

Figure 4 Ego network of food sharing among households of villages A (low OV infection, 4A) and B (high OV infection, 4B), nodes indicate by shapes "Square": Positive households "Triangle": negative households "Circle": households with no fecal examination results, and "Arrow": food sharing links between households.

of sharing-in but significantly different in the number of sharing-out. So, the numbers for giving or selling foods to other households in low infection village were higher than in the village with high infection. Attributing data, such as *O. viverrini* infection and eating raw fish experiences, were taken into account. The prevalence of eating raw fish was declined from last 10 years in both villages, and when simulated the network of food sharing with the history of raw fish eating, the pattern of sharing food was assumed stable. This study demonstrated the different range of relationships for sharing food between low and high *O. viverrini* infection villages which illustrated the influences of sociocultural relationships among people in particular communities for disease transmission (Phimpraphai et al., 2017).

8. PERSPECTIVES

Building effective *O. viverrini* prevention and control programs and integrated approach between multidiscipline and multisectors has to be considered. The 'bottom-up' policies that engage community collaboration and ownership are crucial for program sustainability. These three community-based studies revealed the effectiveness of local stakeholder's participation which is similar to the study by Sripa et al. (2015) which revealed that community public health volunteers are one of the best intermediaries for local information gathering and intervention actions. So, using participatory approach, key informants should be identified and initially participate in the process of questionnaire development to ensure the solid understanding of the study objectives and relevant implementations (Saenna et al., 2017).

The studies also confirmed how deep eating raw fish and sharing food had been embedded in Thai Isaan culture in which rice-fish culture and social connections remain in the villages. The transition from traditional Isaan food-eating practices to more westernized food consumption pattern was found to be according to the availability of modern convenient foods in the village markets or shops (Kim et al., 2017). Lawa Lake region has also become a peri-urban community with the mix of decreasing traditional Isaan cultural practices concurrently happened with the rapidly growing social, economic and ecological changes of modernizing Thailand (Sripa et al., 2015). The trends of eating raw fish had declined among villages located around Lawa Lake area because of the preference of the younger generations for modern western food and the successful of massive liver fluke control campaigns (Kim et al., 2017; Phimpraphai et al., 2017). However, raw fish consumption was still common and had not disappeared in particular groups of people, either men who preferred eating *Koi-pla* with drinking alcohol during social gathering events or elderly people who ate *Pla-som* on a daily meal (Kim et al., 2017). So, health education using community engagement strategies to gain shared issue understanding and collaborative action in intervention program design, implementation and evaluation should be valued for improved local problem analyses, change of community and sustainable health outcomes (Kim et al., 2017).

SNA was used to illustrate the network of fish distribution based on different definitions of nodes and links. Partial network analysis had been carried out among the fishermen, middlemen, fishmongers and *Pla-som* makers to explain the connection of each important actor along the local freshwater fish market chain (Kim et al., 2017). To demonstrate the networks of eating raw fish and fish-based meal sharing among households, the partial networks were calculated by using GIS to compare two high *O. viverrini* prevalence villages with distinct landscape patterns (Saenna et al., 2017). Ego network was also used to explain the different number of food sharing and change of eating raw fish experience between two villages with high and low *O. viverrini* infection (Phimpraphai et al., 2017). So, network parameters, particularly degree (including in-degree and out-degree) that refers to number and direction of sharing, could be used to define the key actors of risky fish distribution and the raw fish dish sharing network and *O. viverrini* transmission pattern among households in Isaan villages. Study of food-sharing links is particularly valuable in households in the rural areas of Thailand where food sharing is related to several public health problems. SNA reflected food-sharing behaviour and demonstrated different food-sharing patterns. Greater connectivity of food sharing among household was associated with increasing number of types of raw fish consumed as well as the frequency of raw fish consumption (Saenna et al., 2017). However, the outcomes from SNA sometimes did not support the hypothesis that food sharing is the main sociocultural factor affecting the transmission of *O. viverrini* among households. So, the others factors such as duration of village settlement, family background, occupation and lifestyles which can affect the social network pattern have been considered. In addition, mapping of the network did provide evidence to support an effective and targeting control strategy. Target households or actors were emphasized by the number of connections between themselves and others, as well as their roles in raw fish-eating practices. These targets should be recruited as a part of community-based intervention that involves promoting properly cooked food or more thoroughly fermenting fish (Saenna et al., 2017). Identification of individuals with high vulnerability to infection could greatly enhance targeted surveillance and prevention programs. Individuals in strongly connected, socially cohesive communities are more likely to perceive economic and social interests as shared. Consequently they may be more motivated and better organized to pursue collective goals such as building and maintaining effective public health interventions. Social network approaches offer a way to apply analytical insights and computerized techniques to move the focus one step above the individual. These particular techniques can be used to demonstrate the flow of social support, social influence and infectious diseases through populations, as well as among organizations. These considerations also emphasize that infectious disease prevention and control must go beyond biomedical sciences and also incorporate the social sciences, as well as associated disciplines. Integrative and participatory approaches have also been considered to encourage authentic stakeholder engagement in developing, implementing and evaluating interventions on liver fluke and other food-borne illness control in the Isaan region.

ACKNOWLEDGEMENTS

The authors gratefully acknowledge the support of Assistant Professor Dr. Chaithep Poolkhet from Kasetsart University for necessary advice. B.S. is a TRF senior research scholar (RTA 5680006). Special thanks go to Dr. Michel de Garine-Wichatitsky from CIRAD and Kasetsart University for editing the manuscript.

REFERENCES

Andrews, R.H., Sithithaworn, P., Petney, T.N., 2008. *Opisthorchis viverrini*: an underestimated parasite in world health. Trends Parasitol. 24, 497–501.

Bansal, S., Read, J., Pourbohloul, B., Meyers, L.A., 2010. The dynamic nature of contact networks in infectious disease epidemiology. J. Biol. Dynam. 4, 478–489.

Bates, S.J., Trostle, J., Cevallos, W.T., Hubbard, A., Eisenberg, J.N.S., 2007. Relating diarrheal disease to social networks and the geographic configuration of communities in rural Ecuador. Am. J. Epidemiol. 166, 1088–1095.

Bell, D.C., Atkinson, J.S., Carlson, J.W., 1999. Centrality measures for disease transmission networks. Soc. Netw. 21, 1–21.

Berkman, L.F., Glass, T., Brissette, I., Seeman, T.E., 2000. From social integration to health: durkheim in the new millennium. Soc. Sci. Med. 51, 843–857.

Berkman, L.F., Glass, T., 2000. Social integration, social networks, social support, and health. In: Berkman, L.F., Kawarchi, I. (Eds.), Social Epidemiology. Oxford University Press, New York, pp. 137–173.

Borgatti, S.P., Everett, M.G., Freeman, L.C., 2002. Ucinet 6 for Windows: Software for Social Network Analysis, vol. 6. Analytic Technologies, Harvard, MA.

Borgatti, S.P., Mehra, A., Brass, D.J., Labianca, G., 2009. Network analysis in the social sciences. Science 323, 892–895.

Case, A., Paxson, C., 2002. Parental behavior and child health. Health Aff. 21, 164–178.

Case, A., Lubotsky, D., Paxson, C., 2002. Economic status and health in childhood: the origins of the gradient. Am. Econ. Rev. 92, 1308–1334.

Chai, J.Y., Shin, E.H., Lee, S.H., Rim, H.J., 2009. Foodborne intestinal flukes in Southeast Asia. Kor. J. Parasitol. 47, S69–S102.

Christakis, N.A., Fowler, J.H., 2007. The spread of obesity in a large social network over 32 years. N. Engl. J. Med. 357, 370–379.

Christley, R.M., Pinchbeck, G.L., Bowers, R.G., Clancy, D., French, N.P., Bennett, R., Turner, J., 2005. Infection in social networks: using network analysis to identify high-risk individuals. Am. J. Epidemiol. 162, 1024–1031.

Cohen, L., Manion, L., Morrison, K., 2007. Research Methods in Education, sixth ed. Routledge, London.

Costello, E.J., Compton, S.N., Keeler, G., Angold, A., 2003. Relationships between poverty and psychopathology. JAMA 290, 2023–2029.

Corner, L.A., Pfeiffer, D., Morris, R.S., 2003. Social-network analysis of *Mycobacterium bovis*-transmission among captive brushtail possums (*Trichosurus vulpecula*). Prev. Vet. Med. 59, 147–167.

Cutler, D.M., Lleras-Muney, A., 2006. Education and health: evaluating theories and evidence. In: NBER Working Paper 12352.

Delormier, T., Frohlich, K.L., Potvin, L., 2009. Food and eating as social practice - understanding eating patterns as social phenomena and implications for public health. Sociol. Health Illn. 31, 215–228.

Doherthy, I.A., Padian, N.S., Marlow, C., Aral, S.O., 2005. Determinants and consequences of sexual networks as they affect the spread of sexually transmitted infections. J. Infect. Dis. 191, S42–S54.

Dubé, C., Ribble, C., Kelton, D., McNab, B., 2011. Introduction to network analysis and its implications for animal disease modelling. Rev. Sci. Tech. 30, 425–436.

Fieldhouse, P., 1995. Social functions of food. In: Food and Nutrition. Springer, Boston, MA.

Friedman, S., Kottiri, B., Neaigus, A., Curtis, R., Vermund, S., Des Jarlais, D.C., 2000. Network-related mechanisms may help explain long-term HIV-1 seroprevalence levels that remain high but do not approach population-group saturation. Am. J. Epidemiol. 152, 913–922.

Friend, R., 2007. Securing Sustainable Livelihoods through Wise Use of Wetland Resources: Reflections on the Experience of the Mekong Wetlands Biodiversity and Sustainable Use Program (MWBP). MWBP, Vientianne. Lao PDR.

Granovetter, M.S., 1973. The strength of weak ties. Am. J. Sociol. 78, 1360–1380.

Grundy-Warr, C., Andrews, R.H., Sithithaworn, P., Petney, T.N., Sripa, B., Laithavewat, L.Z., 2012. Raw attitudes, wetland cultures, life-cycles: socio-cultural dynamics relating to *Opisthorchis viverrini* in the Mekong Basin. Parasitol. Int. 61, 65–70.

Hawe, P., Webster, C., Shiell, A., 2004. A glossary of terms for navigating the field of social network analysis. J. Epidemiol. Community Health 58, 971–975.

Hernandez, L.M., Blazer, D.G., 2006. Genes, behavior, and the social environment: moving beyond the nature/nurture debate. In: Institute of Medicine (US) Committee on Assessing Interactions Among Social, Behavioral, and Genetic Factors in Health. National Academies Press, Washington, DC.

Higgs, S., 2015. Social norms and their influence on eating behaviours. Appetite 86, 38–44.

Higgs, S., Thomas, J., 2016. Social influences on eating. Curr. Opin. Behav. Sci. 9, 1–6.

House, J.S., Landis, K.R., Umberson, D., 1988. Social relationships and health. Sciences 241, 540–545.

Hung, N.M., Madsen, H., Fried, B., 2013. Global status of fish-borne zoonotic trematodiasis in humans. Acta Parasitol. 58, 231–258.

Jongsuksuntigul, P., Imsomboon, T., 2003. Opisthorchiasis control in Thailand. Acta Trop. 88, 229–232.

Kaewpitoon, N., Kaewpitoon, S.J., Pengsa, P., 2008. Opisthorchiasis in Thailand: review and current status. World J. Gastroenterol. 14, 2297–2302.

Kao, R.R., 2002. The role of mathematical modelling in the control of the 2001 FMD epidemic in the UK. Trends Microbiol. 10, 279–286.

Kawachi, I., Kim, D., Coutts, A., Subramanian, S., 2004. Commentary: reconciling the three accounts of social capital. Int. J. Epidemiol. 33, 682–690.

Kim, C.S., Smith, J.F., Suwannatrai, A., Echaubard, P., Wilcox, B., Kaewkes, Sitthithaworn, P., Sripa, B., 2017. Role of socio-cultural and economic factors in cyprinid fish distribution networks and consumption in Lawa Lake region, Northeast Thailand: novel perspectives on *Opisthorchis viverrini* transmission dynamics. Acta Trop. 170, 85–94.

Klovdahl, A.S., Graviss, E.A., Yaganehdoost, A., Ross, M.W., Wanger, A., Adams, G.J., Musser, J.M., 2001. Networks and tuberculosis: an undetected community outbreak involving public places. Soc. Sci. Med. 52, 681–694.
Leavitt, H.J., 1951. Some effects of certain communication patterns on group performance. J. Abnorm. Psychol. 46, 38–50.
Liljeros, F., Edling, C., Amaral, L.A., Stanley, H.E., Aberg, Y., 2001. The web of human sexual contacts. Nature 411, 907–908.
Lima dos Santos, C.A.M., Howgate, P., 2011. Fishborne zoonotic parasites and aquaculture: a review. Aquaculture 318, 253–261.
Lloyd, A.L., May, R.M., 2001. How viruses spread among computers and people. Science 411, 907–908.
Lowy, F.D., Miller, M., 2002. New methods to investigate infectious disease transmission and pathogenesis-*Staphylococcus aureus* disease in drug users. Lancet Infect. Dis. 2, 605–612.
Ma, G., 2015. Food, eating behavior, and culture in Chinease society. J. Ethn. Foods 2, 195–199.
Mansley, L.M., Dunlop, P.J., Whiteside, S.M., Smith, R.G., 2003. Early dissemination of food-and-mouth disease virus through sheep marketing in February 2001. Vet. Rec. 153, 43–50.
Martínez-López, B., Perez, A.M., Sánchez-Vizcaíno, J.M., 2009. Social network analysis. Review of general concepts and use in preventive veterinary medicine. Transbound. Emerg. Dis. 56, 109–120.
Meyers, L.A., Newman, M.E.J., Martin, M., Schrag, S., 2003. Applying network theory to epidemics: control measures for *Mycoplasma pneumonia* outbreaks. Emerg. Infect. Dis. 9, 204–210.
Milgram, S., 1967. The small world problem. Psychol. Today 1, 61–67.
Moreno, J.L., 1934. Who Shall Survive? Nervous and Mental Disease Publishing Company, Washington, DC.
Mutsuyama, A., 2003. Traditional Dietary Culture of Southeast Asia (Its formation and pedigree, Translated by Tomomatsu, A. Kegan Paul, London - New York – Bahrain).
Na Mahasarakarm, O.P., 2007. An Introduction to the Mekong Fisheries of Thailand. Mekong Development Series No. 5. Mekong River Commission, Vientiane, Lao PDR.
Newman, M.E.J., 2002. Spread of epidemic disease on networks. Phys. Rev. E. Stat. Nonlin. Soft Matter Phys. 66, 016128.
Nordström, K., Görman, Á.U., Coff, C., Jönsson, H., Nordenfelt, L., 2013. Food and health: individual, cultural, or scientific matters. Genes Nutr. 8, 357–363.
Oh, A., Erinosho, T., Dunton, G., Perna, F.M., Berrigan, D., 2014. Cross-sectional examination of physical and social contexts of episodes of eating and drinking in a national sample of US adults. Publ. Health Nutr. 17, 2721–2729.
Pahl-Wostl, C., Mostert, E., Tàbara, D., 2008. The growing importance of social learning in water resources management and sustainability science. Ecol. Soc. 13, 24.
Peiris, J.S., Yuen, K.Y., Osterhaus, A.D., Stöhr, K., 2003. The severe acute respiratory syndrome. N. Engl. J. Med. 349, 2431–2441.
Phimpraphai, W., Tangkawattana, S., Sereerak, P., Kasemsuwan, S., Sripa, B., 2017. Social network analysis of food sharing among households in opisthorchiasis endemic villages of Lawa Lake, Thailand. Acta Trop. 169, 150–156.
Pool, I.S., Kochen, M., 1978. Contacts and influence. Soc. Netw. 179, 5–51.
R Core Team, 2013. R: A Language and Environment for Statistical Computing. R Foudation for Statistical Computing, Vienna, Australia. http://www.R-project.org/.
Riolo, C.S., Koopman, J.S., Chick, S.E., 2001. Methods and measures for the description of epidemiologic contact networks. J. Urban Health 78, 446–457.

Rothenberg, R., Baldwin, J., Trotter, R., Muth, S., 2001. The risk environment for HIV transmission: results from the Atlanta and Flagstaff network studies. J. Urban Health 78, 419–432.

Saenna, P., Hurst, C., Echaubard, P., Wilcox, B.A., Sripa, B., 2017. Fish sharing as a risk factor for *Opisthorchis viverrini* infection: evidence from two villages in north-eastern Thailand. Infect. Dis. Poverty 6, 66.

Sattenspiel, L., Mobarry, A., Herring, D.A., 2000. Modeling the influence of settlement structure on the spread of influenza among communities. Am. J. Hum. Biol. 12, 736–748.

Sayasone, S., Odermatt, P., Phoumindr, N., Vongsaravane, X., Sensombath, V., Phetsouvanh, R., Choulamany, X., Strobel, M., 2007. Epidemiology of *Opisthorchis viverrini* in a rural district of southern Lao PDR. Trans. R. Soc. Trop. Med. Hyg. 101, 40–47.

SEAFDEC, 2017. Southeast Asian State of Fisheries and Aquaculture 2017. Southeast Asian Fisheries Development Center, Bangkok, Thailand, 167 pp.

Shin, H.R., Oh, J.K., Masuyer, E., Curado, M.P., Bouvard, V., Fang, Y.Y., Wiangnon, S., Sripa, B., Hong, S.T., 2010. Epidemiology of cholangiocarcinoma: an update focusing on risk factors. Cancer Sci. 101, 579–585.

Sithithaworn, P., Haswell-Elkins, M., 2003. Epidemiology of *Opisthorchis viverrini*. Acta Trop. 88, 187–194.

Sithithaworn, P., Andrews, R.H., Nguyen, V.D., Wongsaroj, T., Sinuon, M., Odermatt, P., Sripa, B., 2012. The current status of opisthorchiasis and clonorchiasis in the Mekong Basin. Parasitol. Int. 61, 10–16.

Sociocultural Factors. Retrieved from http://psychology.wikia.com/wiki/Sociocultural_factors.

Sriamporn, S., Pisani, P., Pipitgool, V., Suwanrungruang, K., Kamsa-ard, S., Parkin, D.M., 2004. Prevalence of *Opisthorchis viverrini* infection and incidence of cholangiocarcinoma in Khon Kaen, Northeast Thailand. Trop. Med. Int. Health 9, 588–594.

Sripa, B., Pairojkul, C., 2014. Cholangiocarcinoma: lessons from Thailand. Curr. Opin. Gastroenterol. 24, 349–356.

Sripa, B., Sithithaworn, P., Sirisinha, S., 2003. *Opisthorchis viverrini* and opisthorchiasis. The 21st century review. Acta Trop. 88, 169–170.

Sripa, B., Kaewkes, S., Intapan, P.M., Maleewong, W., Brindley, P.J., 2010. Food-borne trematodiases in Southeast Asia epidemiology, pathology, clinical manifestation and control. Adv. Parasitol. 72, 305–350.

Sripa, B., Bethony, J.M., Sithithaworn, P., Kaewkes, S., Mairiang, E., Loukas, A., Brindley, P.J., 2011. Opisthorchiasis and *Opisthorchis*-associated cholangiocarcinoma in Thailand and Laos. Acta Trop. 120, 158–168.

Sripa, B., Tangkawattana, S., Laha, T., Kaewkes, S., Mallory, F.F., Smith, J.F., Wilcox, B.A., 2015. Toward integrated opisthorchiasis control in northeast Thailand: the Lawa project. Acta Trop. 141, 361–367.

Sriraj, P., Boonmars, T., Aukkanimart, R., Songsri, J., Sripan, P., Ratanasuwan, P., Boonjaraspinyo, S., Wongchalee, N., Laummaunwai, P., 2016. A combination of liver fluke infection and traditional northeastern Thai foods associated with cholangiocarcinoma development. Parasitol. Res. 115, 3843–3852.

Trostle, J.A., Hubbard, A., Scott, J., Cevallos, W., Bates, S.J., Eisenberg, J.N.S., 2008. Raising the level of analysis of food-borne outbreaks: food-sharing networks in rural coastal Ecuador. Epidemiology 19, 384–390.

Vonghachack, Y., Odermatt, P., Taisayyavong, K., Phounsavath, S., Akkhavong, K., Sayasone, S., 2017. Transmission of *Opisthorchis viverrini, Schistosoma mekongi* and soil-transmitted helminthes on the Mekong Islands, Southern Lao PDR. Infect. Dis. Poverty 6, 131.

Wasserman, S., Faust, K., 1994. Social Network Analysis: Methods and Application. Cambridge University Press.
Watts, D.J., Strogatz, S.H., 1998. Collective dynamics of 'small-world' networks. Nature 393, 440–442.
Wellman, B., Salaff, J., Dimitrova, D., Garton, L., Gulia, M., Haythornthwaite, C., 1996. Computer networks as social networks: collaborative work, telework, and virtual community. Annu. Rev. Sociol. 22, 213–238.
World Health Organization, 1995. Control of foodborne trematode infections. WHO Tech. Rep. Ser. 849, 1–157.
Xayaseng, V., Phongluxa, K., van Eeuwijk, P., Akkhavong, K., Odermatt, P., 2013. Raw fish consumption in liver fluke endemic areas in rural Southern Laos. Acta Trop. 127, 105–111.
Zelner, J.L., Trostle, J., Goldstick, J.E., Cevallos, W., House, J.S., Eisenberg, J.N.S., 2012. Social connectedness and disease transmission: Social organization, cohesion, village context, and infection risk in rural Ecuador. Am. J. Publ. Health 102, 2233–2239.
Ziegler, A.D., Andrews, R.H., Grundy-Warr, C., Sithithaworn, P., Petney, T.N., 2011. Fighting liver flukes with food safety education. Science 331, 282–283.

CHAPTER FIVE

Opisthorchis viverrini Draft Genome — Biomedical Implications and Future Avenues

Neil D. Young[1] and Robin B. Gasser
The University of Melbourne, Parkville, VIC, Australia
[1]Corresponding author: E-mails: nyoung@unimelb.edu.au

Contents

Abstract

Opisthorchiasis is a neglected tropical disease of major proportion, caused by the carcinogenic, Asian liver fluke, *Opisthorchis viverrini*. This hepatobiliary disease is known to be associated with malignant cancer (cholangiocarcinoma, CCA) and affects millions of people in Southeast Asia. No vaccine is available, and only one drug (praziquantel) is routinely employed against the parasite. Despite technological advances, little is known about the molecular biology of the fluke itself and the disease complex that it causes in humans. The advent of high-throughput nucleic acid sequencing and bioinformatic technologies is enabling researchers to gain global insights into the molecular pathways and processes in parasites. The principal aims of this chapter are to (1) review molecular research of *O. viverrini* and opisthorchiasis; (2) provide an account of recent advances in the sequencing and characterization of the genome and transcriptomes of *O. viverrini*; (3) describe the complex life of this worm in the biliary system of the

Advances in Parasitology, Volume 101
ISSN 0065-308X
https://doi.org/10.1016/bs.apar.2018.05.005

definitive (human) host and how the fluke interacts with this host and causes disease at the molecular level; (4) discuss the implications of systems biological research and (5) consider how progress in genomics and informatics might enable explorations of *O. viverrini* and related worms and the discovery of new interventions against opisthorchiasis and CCA.

1. INTRODUCTION

Liver flukes (Class Trematoda: Digenea) are particularly important food-borne worms of humans (Keiser and Utzinger, 2009). *Opisthorchis viverrini* (Opisthorchiidae) causes opisthorchiasis, a disease that has a substantial public health impact in countries of Southeast Asia, including Thailand, Lao People's Democratic Republic (PDR), Cambodia and Vietnam (Sripa et al., 2007). Infection and disease by this parasite affect tens of millions of people, and ~80 million people are at risk of infection (Keiser and Utzinger, 2009; Sripa et al., 2007). Despite control efforts, prevalence can be as high as 70% in some countries such as Thailand (Sithithaworn and Haswell-Elkins, 2003). Chronic *O. viverrini* infection is associated with cholangitis and problems in the liver, most importantly, malignant bile duct cancer (i.e., cholangiocarcinoma, CCA) (Shin et al., 2010; Sripa et al., 2007). In principle, although infection can be controlled if people do not consume raw or inadequately cooked cyprinid fish, opisthorchiasis persists in endemic regions (Keiser and Utzinger, 2009; Sripa et al., 2007), and control relies on treating infected people with the anthelmintic praziquantel. No vaccine is available, and immunogenicity of the fluke is low in humans (Wongratanacheewin et al., 2003), such that they frequently become reinfected (Sithithaworn and Haswell-Elkins, 2003). Although there has been a focus on developing alternative control methods (Keiser et al., 2006, 2009; Keiser et al., 2013), novel flukicides and vaccines are not yet quite 'on the horizon'.

This chapter reviews mainly molecular research of *O. viverrini* and opisthorchiasis. It summarizes recent advances in the sequencing and characterization of the nuclear genome and transcriptomes of *O. viverrini*, discusses the complex life of this worm in the biliary system of the definitive (human) host and how the fluke interacts with this host and causes disease at the molecular level, and describes the implications of recent research advances. The chapter concludes by considering how advances in genomics and informatics might facilitate research of *O. viverrini* and related worms, and the discovery of new interventions against opisthorchiasis and associated CCA.

2. BIOLOGY AND PATHOGENESIS OF DISEASE AND ASSOCIATED CANCER

The mature, hermaphroditic flukes live inside the bile ducts of the definitive host, such as a human, dog or cat. Here, the flukes produce fertilized eggs by sexual reproduction. These eggs are released into the bile and pass, via the gall bladder and the common bile duct, to the chyme in the duodenum, after which they are shed in the faeces into the environment. Eggs are ingested by the first intermediate host — an aquatic snail (order Mesogastropoda) — where the fluke undergoes asexual reproduction. Individual miracidia hatch from individual eggs in the digestive tract of the snail, penetrate its intestinal wall and then transform by metamorphosis into sporocysts. Rediae and cercariae propagate via polyembryony, and cercariae ultimately leave the snail and invade the second intermediate host — a freshwater cyprinid fish. The cercariae penetrate the epidermis and encyst as metacercariae in the musculature and tissues of this host. The cycle completes when the infective metacercarial stage is ingested by a mammalian (piscivorous) definitive host. Metacercariae excyst in the small intestine, an immature fluke is released that migrates via the ampulla of Vater to the extra- or intra-hepatic bile ducts, feeds on bile constituents and matures into an adult fluke within about a month and produces eggs via sexual reproduction.

Infection with *O. viverrini* is prevalent in geographical regions where raw cyprinid fish is a staple of the diet of people (Kurathong et al., 1987). The fluke can live for decades in extra- and intra-hepatic bile ducts and cause chronic cholangitis, leading to periductal fibrosis, cholecystitis and cholelithiasis, obstructive jaundice, hepatomegaly and/or periductal fibrosis of the portal system (Choi et al., 2004; Harinasuta et al., 1984; Hou, 1955; Mairiang et al., 1992; Min, 1984; Sithithaworn et al., 1994). Crucially, chronic infection induces a malignant form of bile duct cancer, i.e., CCA (Haswell-Elkins et al., 1994; Lee et al., 1993; Lim et al., 2006; Thamavit et al., 1978), such that *O. viverrini* is recognized by the International Agency for Research on Cancer as a Group 1 carcinogen (Bouvard et al., 2009).

3. NEED FOR MOLECULAR INVESTIGATIONS

There is no effective drug against CCA, such that prevention must include treatment of *O. viverrini* infection/opisthorchiasis. Although prevention of opisthorchiasis could be achieved by persuading people (via education programs) to consume cooked fish, the cultural custom in endemic areas of consuming raw, undercooked or freshly pickled fish persists (Shin et al., 2010; Sripa and Pairojkul, 2008; Sripa et al., 2015). Thus, opisthorchiasis control still relies on treatment with praziquantel. Despite the apparent efficacy of this drug against the adult worm, humans in *O. viverrini*–endemic regions become readily reinfected, as repeated *O. viverrini* infections do not induce protective immunity in humans (Sripa et al., 2007) and repeated treatment risks the induction of resistance to praziquantel in the fluke (Fallon and Doenhoff, 1994). Therefore, there is a need to search for alternative approaches and targets for prevention and control. In our opinion, this search could be built on a sound understanding of the molecular biology of *O. viverrini*, the crosstalk between the fluke and its definitive hosts, and the diseases that the fluke cause or induce. Indeed, global investigations of the genomes, transcriptomes and proteomes of *O. viverrini* isolates should assist in elucidating the molecular pathogenesis of opisthorchiasis and oncogenesis of CAA and could provide a foundation for identifying targets for the design of new interventions.

Prior to the availability of the genome of *O. viverrini*, a number of molecular studies (e.g., Cancela et al., 2010; Gourbal et al., 2008; Ju et al., 2009; Kim et al., 2008; Laha et al., 2007; Mulvenna et al., 2010; Nagaraj et al., 2007) indicated exciting prospects to explore biochemical, physiological and biological pathways in liver flukes and to predict and prioritize novel drug targets. In particular, an earlier study was able to assemble ~2000 short transcripts from adult *O. viverrini* using a Sanger sequencing approach (Laha et al., 2007). However, in the early to mid-2000s, the amount of data had been inadequate to undertake or support global molecular investigations of this fluke, limiting our ability to comprehensively understand its biology and to work towards new interventions.

The advent of high-throughput sequencing approaches provided a path to obtain a more global perspective of the molecular landscape of *O. viverrini*. In the first instance, Young et al. (2010a) used a semiautomated bioinformatic platform to assemble and annotate a nonredundant transcriptome for adult *O. viverrini*. Here, normalized, complementary nucleotide sequences (n = 500,000) were obtained by 454 sequencing (Roche), assembled de novo and then clustered. Redundancy was reduced by clustering sequences – if their predicted proteins aligned over 60% of their length and shared 95% amino acid residue identity. By interrogating nucleotide and conceptually translated amino acid sequences against public and nonredundant databases (Young et al., 2010b), homologues were identified in various eukaryotes. Proteins were further annotated based on sequence homology across conserved protein domains identified using the program InterProScan (Zdobnov and Apweiler, 2001). Following annotation, individual molecules were characterized and categorized according to their presence in eukaryotic biochemical pathways, cellular location (intracellular, extracellular or transmembrane) and gene ontology (GO). At the time, 85% of the sequence data were new, expanding public databases (Cho et al., 2006, 2008; Jex et al., 2012; Ju et al., 2009; Laha et al., 2007; Young et al., 2010a) and providing critical resources to research groups working on molecular aspects of *O. viverrini*.

Comparative analyses showed that opisthorchiids share more (29%–31%) protein sequence homology with fasciolids than with blood flukes (schistosomes). The proteins encoded by adult *O. viverrini* mapped to metabolic, genetic and environmental data/information and cellular processing pathways inferred to be conserved amongst eukaryotes, whereas 90% of predicted proteins lacked sufficient sequence homology to be mapped to any known pathways. This observation provides prospects to explore entirely new biological processes and pathways unique to carcinogenic parasites or to parasite—host crosstalk, some of which might represent intervention targets that are specific to *O. viverrini*. Interesting and somewhat unexpected was the finding that more *O. viverrini* proteins were homologous to proteins of mammals (20%) than of nematodes (15%), in agreement with previous studies of other parasitic flukes (schistosomes) (Berriman et al., 2009; *Schistosoma japonicum* Genome Sequencing and Functional Analysis Consortium, 2009). This finding might suggest that there are molecules in *O. viverrini* that are able to regulate host responses via immunomolecular, biochemical or signalling pathways (Jex et al., 2012; Young et al., 2010a,b).

4. BUILDING GENOMIC RESOURCES AS A BASIS TO UNDERSTAND *OPISTHORCHIS VIVERRINI* AND ASSOCIATED DISEASES AT THE MOLECULAR LEVEL

Advances of knowledge through the use of next-generation transcriptomics were significant, but it became necessary to explore the genome to underpin meaningful and deep molecular investigations. To address this aspect, in 2014, we defined a draft genome for *O. viverrini* (see Young et al., 2014) and undertook detailed analyses of the molecular biology of the worm using transcriptomic and predicted proteomic data sets. Approximately 80 Gb of short-read sequence data (134-fold genome coverage) were produced from seven genomic DNA libraries (insert sizes: 170 bp to

20 kb) built from adult worms of *O. viverrini* ($n = 25$). Frequency distributions (17-mer) revealed low sequence heterozygosity among and within short paired-end libraries. The draft genome was 634.5 Mb in size (N50 = 1,323,951 bp; longest scaffold 9,657,388 bp), with 86% of 248 core essential genes detected, consistent with a well-assembled trematode genome (Berriman et al., 2009; Cwiklinski et al., 2015; *Schistosoma japonicum* Genome Sequencing and Functional Analysis Consortium, 2009; Young et al., 2012). The average GC content of all genomic scaffolds (44%) was similar to that of *Clonorchis sinensis* (see Huang et al., 2013) and *Fasciola hepatica* (see Cwiklinski et al., 2015).

The draft genome assembly of *O. viverrini* was compared with other representative fluke genomes and showed unique genomic features (Young et al., 2014). An initial comparison with *C. sinensis* showed marked structural variation between the two genomes. Only 22% of *O. viverrini* scaffolds aligned to 26% of the *C. sinensis* scaffolds at the nucleotide level, and there was limited genomic synteny between these two species (based on scaffolds of >100 kb). Substantial divergence was also seen when *O. viverrini* was compared with *Schistosoma mansoni*. In total, 308 *O. viverrini* scaffolds with sequence similarity to chromosome 1 of *S. mansoni* (65,476,681 bp) aligned in 107 syntenic blocks, covering 35% of the *O. viverrini* and 8% of the *S. mansoni* scaffolds. This lack of synteny likely attributes to nucleotide and karyotypic differences among these fluke species. Published studies report 12 chromosomes for *O. viverrini* (Kaewkong et al., 2012), 14 or 58 for *C. sinensis* (see Park et al., 2000) and 8 for human schistosomes (Short and Menzel, 1960). Thus, the *O. viverrini* draft genome is unique when compared with other published fluke genomes (Berriman et al., 2009; Cwiklinski et al., 2015; Huang et al., 2013; *Schistosoma japonicum* Genome Sequencing and Functional Analysis Consortium, 2009; Young et al., 2012).

Protein-coding genes, noncoding RNAs and repeat content were then identified in the *O. viverrini* genome. A total of 16,379 protein-encoding genes were inferred from the draft genome using RNA-seq data sets and sequence similarity to protein-encoding genes of *C. sinensis* (see Huang et al., 2013) and blood flukes (Berriman et al., 2009; *Schistosoma japonicum* Genome Sequencing and Functional Analysis Consortium, 2009; Young et al., 2012) as evidence. The majority of genes (87%) were supported by published data (Jex et al., 2012; Young et al., 2010a) for adults and juveniles of *O. viverrini*, and nearly all de novo—assembled transcripts (>99%) mapped to the draft genome. The total number of genes, the proportion of coding regions (3.4%) and the mean total gene length (18,231 bp), intron length

(3531 bp), exon length (254 bp) and mean number of exons per gene (5.8) were similar to those of *C. sinensis* (see Huang et al., 2013) but distinct from blood flukes. The *O. viverrini* draft genome was shown to exhibit marked intronic expansions compared with other flukes (mean 2.8–3.5 kb and equating to a 1.2- to 1.5-fold increase in intron length) (Young et al., 2014).

Noncoding RNAs (n = 806), such as tRNAs (189) and conserved microRNAs (178), were inferred for *O. viverrini*. There was a significant relationship between the frequency of encoded tRNA copies and amino acid usage. The amino acid composition of translated protein-coding domains in *O. viverrini* was akin to that of *F. hepatica* (see Cancela et al., 2010) and of cestodes (McSorley et al., 2013; Tsai et al., 2013), indicating that GC-rich domains in the *O. viverrini* draft genome code for valine, proline, glycine, alanine and arginine amino acids more frequently, and for tyrosine, serine, isoleucine and asparagine less frequently than in schistosomes.

Approximately one-third (31%) of the *O. viverrini* genome contains conserved repeats — similar to *C. sinensis* (26%; Huang et al., 2013) but distinct from schistosomes (40%–48%; Berriman et al., 2009; *Schistosoma japonicum* Genome Sequencing and Functional Analysis Consortium, 2009; Young et al., 2012). Almost 62% of repeats predicted were retrotransposons, such as long terminal repeat (LTR)/*Pao* (0.9%), LTR/*Gypsy* (8.4%), LTR (9.7%) and unclassified elements. Long interspersed elements (LINEs), including LINE/*Penelope* (5.2%), LINE/CR1 (8.2%) and LINE/RTE (24%) elements, were predicted.

5. A 'SYSTEMS BIOLOGICAL' LOOK AT THE COMPLEXITY OF THE FLUKE'S LIFE IN THE BILE DUCT

Using advanced bioinformatic pipelines, we integrated genomic, transcriptomic, inferred proteomic data and published information to explore the molecular landscape of *O. viverrini* (see Young et al., 2014). This effort allowed us to address outstanding questions regarding the biology of this fluke in the biliary system (this section), essential pathways in the fluke and those linked to fluke–host crosstalk (Section 5) and genes/gene products that might contribute to CCA development (Section 6).

Our detailed molecular explorations (Young et al., 2014) revealed that *O. viverrini* transcribes genes encoding vast repertoires of enzymes and receptors are likely essential for the absorption, binding and/or conversion of nutrients from bile and the biliary system. Such adaptations likely permit this fluke to effectively use dominant constituents within bile ducts to survive under extremely harsh physiological conditions.

Following ingestion by the human host, *O. viverrini* metacercariae pass through the stomach, excyst in the small intestine and the juvenile flukes migrate to and eventually establish as adults in the biliary ducts (Harinasuta and Harinasuta, 1984). In the biliary system, the fluke is exposed to harsh conditions and needs to grow, develop, reproduce and survive in a nutrient-poor environment. In the initial phase within the host, cysteine and aspartic proteases likely initiate excystment of metacercariae, and ion channels and GPCRs facilitate chemotaxis-mediated metacercarial migration to the biliary system. Once in this system, the fluke is exposed to varying, often low oxygen tension, which is why the fluke expresses a haemoglobin that has a high oxygen-binding capacity (Kiger et al., 1998) and genes associated with both aerobic (such as pyruvate kinase) and anaerobic (including phosphoenolpyruvate carboxykinase) glycolysis (van Hellemond et al., 2003).

Although the fluke can use glycolysis for energy production, bile is usually depleted of glucose (Masyuk et al., 2002). Therefore, flukes in the biliary system transcribe genes encoding enzymes and accessory proteins involved in processing bile constituents through the degradation of proteins (via branched-chain amino acid metabolism) and free lipids (via mitochondrial fatty acid beta-oxidation) (Young et al., 2014). Interestingly, fluke lipases appear to degrade micelles and host cell membranes (Hu et al., 2009), to increase the availability of lipids, carbohydrates and proteins to the worm. The fluke produces other enzymes and accessory proteins that likely further degrade micelles. Such enzymes include cysteine (cathepsins F and B and AEPs), serine (chymotrypsin-like) and aspartic (pepsin-like and cathepsin D) proteases transcribed in juvenile and adult stages (Young et al., 2014). Collectively, these proteases have broad substrate specificity and likely degrade many proteins and lipoprotein complexes. Once degraded, the transfer of free, bile-derived amino acids (likely via amino acid transporters) into the fluke are likely processed for energy production via acetyl CoA.

The fluke is exposed to host immune attack. Therefore, evading this attack is crucial to the fluke's survival, particularly if inflammation follows invasion and cellular damage caused by the fluke. Analyses of transcription data (Young et al., 2014) indicate that excretory/secretory (ES) proteins modulate host immune responses. Genes encoding defence proteins (e.g., T265_13308) are highly transcribed in bile duct stages (Young et al., 2014). Such proteins are hypothesized to mimic cathelicidin (a mammalian host defence peptide) and suppress a Th1 response by averting LPS interaction with Toll-like receptor 4 on macrophages (Robinson et al., 2011). Cathepsin F peptidases from the fluke likely degrade immunoglobulins, including secretory immunoglobulin A (Kang et al., 2010).

The fluke is exposed to bile and noxious substances, such as drugs and metabolites (xenobiotics), toxins and carcinogens (Trauner and Boyer, 2003), which can induce free oxygen radical formation via lipid peroxidation (Lechner et al., 2002). The fluke protects itself using antioxidants, such as glutathione-S-transferases (GSTs), that reduce lipid hydroperoxides (Sharma et al., 2004), detoxify xenobiotic substrates via glutathione conjugation (Trauner and Boyer, 2003) as well as SODs that convert free radicals to hydrogen peroxide. Based on transcription, glutathione-like and peroxiredoxin—peroxidases in the fluke convert hydrogen peroxide to water.

The fluke needs to grow, develop, survive and reproduce in a nutrient-poor environment. Initially, the newly excysted juvenile fluke relies on energy from glycogen granules and lipid droplets stored in the excretory bladder (Orido, 1990). As these energy reserves rapidly deplete, the developing fluke must acquire essential energy and nutrient resources for growth, development and reproduction. The fluke can use lipoproteins (Ginsberg, 1998) as bile is rich in very low—density lipoproteins, low (LDLs), intermediate (IDLs) and high (HDLs), which contain varying proportions of cholesterol, phospholipids, triglycerides and amphipathic proteins (Ginsberg, 1998). Bile is also rich in arachidonic (C20:4) fatty acids, unsaturated linoelaidic (C18:2), long-chain saturated palmitic acid (C16:0) and branched-chain amino acids (Folsch and Wormsley, 1977; Halpern et al., 1993).

As the fluke itself does not directly synthesize cholesterol, it maintains key molecular processes to acquire, transport and process cholesterol from the external lipid-rich environment in the bile duct. This fluke transcribes a scavenger-like receptor (SR-B1), which transports cholesterol from HDLs across plasma membranes (Acton et al., 1996). It uses proteins such as the LDL receptor (LDLR), LDLR-related protein 1 receptor (LRP1) and CD36-like receptors to facilitate LDL, IDL and fatty acid uptake, respectively (Young et al., 2014). In the late lysosome and endosome compartments, cholesterol esters are hydrolyzed via an acid lipase to free cholesterol (Young et al., 2014). It appears that cholesterol is bound to one NPC2-like protein and transported, via the Niemann—Pick C1 and ATP-binding cassette transporter 1 complex (Subramanian and Balch, 2008), to the cytosol. Once bound to oxysterol-binding protein-related proteins, cholesterol is converted to a cholesterol ester by lecithin:cholesterol acyltransferase and then eliminated by an MTTP/PDI-like complex.

The fluke has a large family of novel lipid-binding proteins (n = 25), 15 of which are transcribed in juvenile and adult stages (Young et al., 2014). These proteins contain a conserved MD-2-related lipid-binding domain and are homologous to the human Niemann–Pick C2 protein, which enables intracellular and extracellular sterol transport (Inohara and Nunez, 2002). Interestingly, other eukaryotes only have one copy of this NPC2-like protein (Inohara and Nunez, 2002). The expansion of this protein family likely reflects the need for the fluke to bind and transport sterols and/or lipids, including cholesterol. There was also some evidence that the fluke likely digests cholangiocytes, as juvenile and adult worms transcribe specific sphingomyelin phosphodiesterase and galactosylceramidase/galactocerebrosidase genes. These enzymes in the lysosome are crucial for catabolizing sphingomyelin, which is enriched in cholangiocytes (Amigo et al., 1999).

6. ESSENTIAL PROCESSES OR PATHWAYS IN THE FLUKE AND THOSE LINKED TO FLUKE–HOST 'CROSSTALK'

The 16,379 protein-coding genes were annotated based on sequence similarity to publicly accessible sequence, biological pathway and GO data sets (Young et al., 2014). A total of 9402 genes (57%) could be assigned to 184 biological pathways and conserved components of nucleotide, carbohydrate, lipid and amino acid metabolisms established. In addition to cellular processing components, such as proteins linked to cell turnover, protein transport and catabolism, specific protein classes, such as cytoskeletal proteins (212), ion channels (128), receptors (140) including G protein-coupled receptors (59), kinases (n = 262), glycosyltransferases (84), transcription factors (226) and proteases (386), were defined. The proteases could be divided into five major classes (aspartic, cysteine, metallo-, serine and threonine), with the latter two classes predominating (21%–44%). Multiple copies of secreted aspartic peptidases (e.g., cathepsin D; n = 27 genes), major repertoires of cathepsins B, F and L, as well as C13-like asparaginyl endopeptidases (94 distinct genes) were present, which is consistent with results from previous transcriptomic data sets (Jex et al., 2012; Young et al., 2010a).

Amazingly, 45 distinct proteases are encoded by the top 10% of abundantly transcribed genes in both juvenile and adult stages; these proteases include cathepsins B, D and F, calpain, a tryptase-like serine peptidase and C13-like asparaginyl endopeptidase (Young et al., 2014). Despite the differential transcription between juvenile and adult stages of numerous, novel asparaginyl endopeptidase genes, their functional roles are enigmatic (Young et al., 2014), although it might be that they activate or transprocess other peptidases with Asn residues at the point of cleavage (Dalton and Brindley, 1996). As expected, based on the extensive sets of proteases in *O. viverrini*, a diverse panel of inhibitors of cysteine (cystatins), serine (α-2-macroglobulin and serpin) and aspartic (α-2-macroglobulin) proteases, whose transcripts were enriched in both juvenile an adult fluke stages, was present (Young et al., 2014).

A comparative analysis revealed >8500 homologous proteins among *O. viverrini*, *C. sinensis*, *S. haematobium*, *S. japonicum* and *S. mansoni*, and 3015 genes (18%) that were unique to *O. viverrini* (Young et al., 2014). In total, >2000 orthologous groups were identified in opisthorchiids (*O. viverrini* and *C. sinensis*) that are divergent from those in blood flukes and tapeworms. More than 1900 of them (single-copy, one-to-one orthologues) represent proteins involved in physiological processes and development that are unique to opisthorchiids (Young et al., 2014). For instance, some enzymes appear to function in cell movement (leishmanolysin-like peptidases) or feeding (such as pepsin A and cathepsin D). Others likely participate in the glycosylation of secretory proteins (glycoprotein-N-acetylgalactosamine 3-beta-galactosyltransferases and alpha-1,6-mannosyl-glycoprotein 6-beta-N-acetylglucosaminyltransferases) or in glycosphingolipid biosynthesis (N-acetyllactosaminide beta-1,6-N-acetylglucosaminyl-transferase) (Young et al., 2014). Other proteins include ubiquitin-protein ligases that bind to specific protein substrates, catalyzing the transfer of ubiquitin and signalling protein degradation or protein—protein interactions[27]; enhancers (GATA-binding proteins and SOX 1/2/3/14/21) that regulate sex determination and other gene regulatory networks involved in developmental progression (Spitz and Furlong, 2012); transcription repressors (Krüppel-associated box domain-containing zinc finger proteins) and cytoskeletal microtubule—mediated organelle transport and centrosome assembly (dyneins and kinesins).

ES proteins of *O. viverrini* are likely central to fluke—host crosstalk and/or contribute to the oncogenesis of CCA (Mulvenna et al., 2010). This fluke's secretome is predicted to contain 437 proteins, including key groups of venom allergen-like (VAL) proteins, proteases, heat shock proteins and SODs (Young et al., 2014). ES proteins conspicuous in the juvenile and adult flukes include progranulin (*Ov*-PGRN), two opisthorchiid-specific granulins (*Ov*-GRN1 and *Ov*-GRN2), cysteine peptidases (CatB, AEP), Niemann–Pick C2 (NPC2) proteins, vitelline B, repetin, TP and GST (Young et al., 2014). VAL proteins (McSorley et al., 2013), cathepsin L (Smith et al., 1993) and granulins (Papatpremsiri et al., 2015; Smout et al., 2009) are implicated in parasite–host crosstalk or the pathogenesis of disease.

7. GENES/GENE PRODUCTS THAT MIGHT CONTRIBUTE TO CHOLANGIOCARCINOMA DEVELOPMENT

An exome sequencing study of genomic DNA from humans with known *O. viverrini* infection clearly shows the multifactorial nature of CCA (Ong et al., 2012), with mutations in protooncogenes and tumour suppressor genes. Factors contributing to an accumulation of epigenomic and genomic modifications in cholangiocytes include (1) fibrosis linked to a metastasis-associated protein 1 and an interleukin-6 (IL-6)–regulated inflammatory response to chronic infection (Nair et al., 2011; Sripa et al., 2009); (2) cell damage or death linked to fluke feeding and/or attachment (Sripa et al., 2012); (3) transforming growth factor-beta and/or epidermal growth factor–mediated cell proliferation linked to ES molecules from the fluke (Thuwajit et al., 2006); and (4) exogenous carcinogens, such a nitrosamines present in diets containing fermented fish and pork (Honjo et al., 2005).

A number of processes or mechanisms have been proposed to explain how chronic *O. viverrini* infection contributes to CCA. Cell damage, cell death and inflammation associated with fluke attachment and/or feeding are thought to contribute the pathogenesis of disease. The bile duct epithelium shields underlying cells (such as hepatocytes) from lipid loss by maintaining abundant concentrations of sphingomyelin and cholesterol at the apical membrane of cholangiocytes (Amigo et al., 1999), and deprotonated bile acids by an umbrella of bicarbonate (HCO_3^-) from these cells (Beuers et al., 2010). Epithelial damage can result in the conjugation of bile acids that can increase the expression of inflammatory mediators (such as IL-6 and IL-8), leading to oxidative stress, DNA damage, fibrosis and/or oncogenesis (Baptissart et al., 2013).

Other evidence indicates that *O. viverrini*–released molecules, such as granulin-like growth factors, modulate the proliferation of host cells (Smout et al., 2009). The granulin-like growth factor *Ov*-GRN-1 and a novel single-domain granulin (*Ov*-GRN2), whose genes are transcribed in both juvenile and adult stages (Young et al., 2014), and progranulin (*Ov*-PRGN), all likely contribute to the regulation of cell development and maintenance. Future studies should compare the mitogenic potential of *Ov*-GRN-2 against *Ov*-GRN-1 (Young et al., 2014). As cathepsin D can stimulate the growth of some human cancers, via the proteolytic degradation of growth regulators, growth factor receptors and extracellular matrix components (Benes et al., 2008), other work might assess the oncogenic potential of members of cathepsin D protease group.

8. CONCLUSIONS AND CHALLENGES FOR THE FUTURE

The draft genome and transcriptomes of *O. viverrini* represent a useful starting point for in-depth studies of the biology of this fluke and its pathogenic, including carcinogenic, effects on its human host. Building on previous work, it will be important to realize the promise of genomics and transcriptomics in different areas. In the following are some avenues that we think are worthy of pursuit.

First, it will important to complete the genome. Although the published draft genome and transcriptomes have been excellent resources, they are incomplete. New sequencing technologies, such as long-range Illumina sequencing (Hi C and/or Chicago libraries) or PacBio genome sequencing and assembly of long DNA reads (cf. Berlin et al., 2015; Korhonen et al., 2016; Madoui et al., 2015; Sharon et al., 2015) can achieve better quality genome assemblies than that of the current *O. viverrini* draft genome, although the cost for PacBio sequencing of large, trematode genomes (0.5—1.3 Gb) would be out of reach for most molecular parasitology laboratories. This 'third-generation' sequencing approach has stimulated the design of new computer algorithms for assembly and postassembly, aimed at delivering a 'haploid' genome assembly from 'diploid' sequencing data (Bodily et al., 2015; Safonova et al., 2015). The combined use of new sequencing and assembly techniques should allow the finishing of multiple strains of *O. viverrini* to chromosomal-level contiguity. Such an effort might resolve problems associated with the assembly of multigene families, which cannot be addressed using short sequence read data sets and current data sets (Alkan et al., 2011), but that will be crucial to understanding structural aspects of the genome and many aspects of the biology, epidemiology, physiology and biochemistry of the parasite (as addressed in other chapters of this issue). Importantly, improved genome and associated transcriptomic data sets for *O. viverrini* can be integrated into ParaSite at WormBase (www.wormbase.org), as a public resource, so that all data are readily accessible and usable by, for instance, individual researchers working on small, biologically focused projects and system biologists working on large-scale data mining. Fortunately, the computational substructure of WormBase is such that the bioinformatic tools (enabling data queries, genome browsing, motif searching for noncoding DNA elements, genome-wide paralogies and orthologies) are very flexible (Harris et al., 2014).

Second, with improved genome assemblies, comparative genomics should be pursued. Such analyses should more be feasible, given the expansion in number of trematode genomes (n = 10) that are now available via ParaSite (Harris et al., 2014). Genomic comparisons between both parasitic and nonparasitic trematode species could enable the identification of genes in *O. viverrini* undergoing gene family amplification and being likely linked to parasitism (Jex et al., 2012; Young et al., 2014). However, comparative genomics still faces significant technical challenges when working with draft genomes. To overcome the challenges of comparing draft genomes and assessing them for their applicability for population genetic studies, Wang et al. (2018) established a refined bioinformatic workflow to define high-quality syntenic blocks between genomes of the same species (in this case, *C. sinensis*). Using this workflow, these authors compared draft genomes of *C. sinensis* from Korea and China, revealed a high level of nucleotide similarity within the syntenic regions of the two genomes and identified two variable genes that encode proteins likely to participate in lipid metabolism and linked to the life of *C. sinensis* in the host's biliary system. By comparing syntenic blocks, they assessed nucleotide variability between the genomes and identified synonymous and nonsynonymous mutations in protein-coding genes involved in diverse biological processes. This study emphasized the challenges associated with draft genome comparisons and delivered a practical workflow for the definition of conserved and variable genes to enable an assessment of their utility as population markers (from a fundamental perspective) or as intervention targets (from an applied viewpoint). The evidence of cryptic species within *O. viverrini* (see Petney et al., 2013) highlights the importance of studying and comparing the genomes of geographically distinct isolates of this fluke. Research using selected genetic loci (cf. Petney et al., 2013) seems to suggest that variation in this parasite has been influenced by factors including life cycle, environmental adaptation and climatic change. As it is now feasible to undertake population investigations using large nuclear genomic data sets, produced using high-throughput sequencing technologies and assembled employing advanced bioinformatics (Crellen et al, 2016; Young et al., 2015), explorations of the genetic structures and substructures of *O. viverrini* populations, and of variation in genes or gene products involved in infection and the associated disease complex, seems particularly relevant.

Third, it would be exciting to explore how the *O. viverrini* transcriptome is regulated by noncoding DNA elements to elucidate gene families that are coregulated, such as those whose expression and activity might govern parasitism or disease. This area is likely of major importance to understanding parasitism, because it has been shown that parasitic trematodes secrete miRNAs (e.g., within exosomes) that can exert immunomodulatory effects on the mammalian host (reviewed by Cai et al., 2016; Fromm et al., 2017). There have been some descriptive studies of ncRNA genes in opisthorchiids, including *O. viverrini, O. felineus*, and *C. sinensis* (see Ovchinnikov et al., 2017). For example, a recent study (Ovchinnikov et al., 2017) claims that these three species have very similar miRNA complements, with 53 identical and 2 specific miRNA genes that represent promising candidates for further investigations and as potential biomarkers. Investigating functional aspects of these miRNA complements represents an exciting area for future work. In particular, it would be interesting to establish whether and, if so, which miRNAs of *O. viverrini* associate with biological pathways to trigger the oncogenesis of CCA in humans. While no direct link has yet been made between the parasite and the human host, a number of studies have shown an association of miRNA profiles with CCA and/or subtypes thereof, suggesting an involvement of particular miRNAs in tumour progression and/or metastasis; miRNA signatures associated with *O. viverrini*—induced CCA also show potential as biomarkers for the early detection of this cancer (Chusorn et al., 2013; Peng et al., 2014; Plieskatt et al., 2014, 2015; Silakit et al., 2014, 2017). In spite of this progress, no comprehensive study of *O. viverrini* ncRNAs has been carried out using a complete or near complete parasite genome as a reference. With a high-quality genome in place, it would be possible to scan a large genome (0.8—2 Gb) in a day and to predict ncRNAs using the steadily expanding Rfam database (Nawrocki et al., 2015; http://rfam.xfam.org/), which currently contains 2686 ncRNA families. Therefore, it should be feasible to predict a comprehensive ncRNA set and miRNAs subsets encoded in the *O. viverrini* genome and to explore which of these snRNAs might contribute to parasitism, opisthorchiasis and/or CCA. In addition to this focus, it would also be important to explore the transcription and expression of genes in different tissues and cells in relation to parasitism, infection and disease. What could be done are analyses of sets of cells of the gut, tegument, reproductive system and/or sensory neurons that might regulate *O. viverrini* development or differentiation during infection. Microdissection of small numbers of cells, followed by RNA amplification and RNA-seq is possible and has been achieved in, for example, *S. mansoni* (see Gobert et al., 2009; Nawaratna et al., 2011; Nawaratna et al., 2014). It may even be possible to express a transgene in the cell type of interest and employ transgenic expression to yield specific cell types from dissociated larvae or adults, through cell sorting (e.g., Spencer et al., 2014), provided a robust transgenesis system could be developed.

Fourth, studying gene regulation would benefit enormously from having a robust functional genomics platform for *O. viverrini* that provides reproducible and persistent gene knockdown. Although quite challenging in trematodes, RNA interference (RNAi) has been achieved in *O. viverrini* (see Sripa et al., 2011; Chaiyadet et al., 2017) and particularly in blood flukes (schistosomes) (reviewed by Hagen et al., 2015a,b). For instance, recently, Hagen et al. (2014) established a lentivirus-based transduction system to deliver microRNA-adapted short hairpin RNAs into *S. mansoni* to silence and explore selected protein-encoding genes implicated in disease. Specifically, these workers showed that the knockdown of the *omega-1* gene in *S. mansoni* eggs resulted in a significant decrease in granuloma size in mice. The increased infiltration of dendritic cells, T-helper and B cells, and interstitial macrophages in the lungs of (empty vector-) control mice was consistent with that expected of egg-induced disease and was clearly abrogated by *omega-1* gene knockdown. The specificity and persistence of gene knockdown in *S. mansoni* eggs meant that in vivo experimentation could be conducted to measure immunopathological responses against eggs in mice, which would likely not have been possible using a conventional knockdown. This lentivirus-based approach likely overcomes the limitations of conventional RNAi (via electroporation or soaking), including possible off-target effects of RNAi triggers and a lack of persistent knockdown over a longer period of time required for in vivo experiments in animals). Although double-stranded RNA and siRNA usually achieve gene knockdown for periods of up to 4 weeks in *S. mansoni* in vitro (Correnti et al., 2005; Krautz-Peterson et al., 2010), siRNA does not induce a long-term effect in vivo, when schistosomules are transferred to the definitive host (Krautz-Peterson et al., 2010). Clearly, the integration of a lentivirus encoding an RNAi-trigger expression cassette (Hagen et al., 2014) circumvents this limitation to achieving sustained knockdown in *S. mansoni* in mice. Thus, this gene-silencing system should be adaptable for in vivo functional genomic—phenomic studies of *O. viverrini* in hamsters. It presents an opportunity to reveal host—parasite interactions and pathomechanisms in opisthorchiasis and CCA and could enable the development of novel interventions against this destructive disease complex.

ACKNOWLEDGEMENTS

Funding from the National Health and Medical Research Council (NHMRC) of Australia (R.B.G. and N.D.Y.), the Australian Research Council and Melbourne Water Corporation and the University of Melbourne (BIP) is gratefully acknowledged (R.B.G. and N.D.Y.). N.D.Y. holds an NHMRC Career Development Fellowship. The authors thank all coauthors and collaborators on original research articles cited in this chapter and the Melbourne Bioinformatics Platform for support.

REFERENCES

Acton, S., Rigotti, A., Landschulz, K.T., Xu, S., Hobbs, H.H., Krieger, M., 1996. Identification of scavenger receptor SR-BI as a high density lipoprotein receptor. Science 271, 518—520.

Alkan, C., Sajjadian, S., Eichler, E.E., 2011. Limitations of next-generation genome sequence assembly. Nat. Methods 8, 61e65.

Amigo, L., Mendoza, H., Zanlungo, S., Miquel, J.F., Rigotti, A., Gonzalez, S., Nervi, F., 1999. Enrichment of canalicular membrane with cholesterol and sphingomyelin prevents bile salt-induced hepatic damage. J. Lipid Res. 40, 533—542.

Baptissart, M., Vega, A., Maqdasy, S., Caira, F., Baron, S., Lobaccaro, J.M., Volle, D.H., 2013. Bile acids: from digestion to cancers. Biochimie 95, 504—517.

Benes, P., Vetvicka, V., Fusek, M., 2008. Cathepsin D—many functions of one aspartic protease. Crit. Rev. Oncol. Hematol. 68, 12—28.

Berlin, K., Koren, S., Chin, C.S., Drake, J.P., Landolin, J.M., Phillippy, A.M., 2015. Assembling large genomes with single-molecule sequencing and locality-sensitive hashing. Nat. Biotechnol. 33, 623e630.

Berriman, M., Haas, B.J., LoVerde, P.T., Wilson, R.A., Dillon, G.P., Cerqueira, G.C., Mashiyama, S.T., Al-Lazikani, B., Andrade, L.F., Ashton, P.D., Aslett, M.A., Bartholomeu, D.C., Blandin, G., Caffrey, C.R., Coghlan, A., Coulson, R., Day, T.A., Delcher, A., DeMarco, R., Djikeng, A., Eyre, T., Gamble, J.A., Ghedin, E., Gu, Y., Hertz-Fowler, C., Hirai, H., Hirai, Y., Houston, R., Ivens, A., Johnston, D.A., Lacerda, D., Macedo, C.D., McVeigh, P., Ning, Z., Oliveira, G., Overington, J.P., Parkhill, J., Pertea, M., Pierce, R.J., Protasio, A.V., Quail, M.A., Rajandream, M.A., Rogers, J., Sajid, M., Salzberg, S.L., Stanke, M., Tivey, A.R., White, O., Williams, D.L., Wortman, J., Wu, W., Zamanian, M., Zerlotini, A., Fraser-Liggett, C.M., Barrell, B.G., El-Sayed, N.M., 2009. The genome of the blood fluke *Schistosoma mansoni*. Nature 460, 352—358.

Beuers, U., Hohenester, S., de Buy Wenniger, L.J., Kremer, A.E., Jansen, P.L., Elferink, R.P., 2010. The biliary HCO(3)(-) umbrella: a unifying hypothesis on pathogenetic and therapeutic aspects of fibrosing cholangiopathies. Hepatology 52, 1489—1496.

Bodily, P.M., Fujimoto, M., Ortega, C., Okuda, N., Price, J.C., Clement, M.J., Snell, Q., 2015. Heterozygous genome assembly via binary classification of homologous sequence. BMC Bioinform 16 (Suppl. 7), S5.

Bouvard, V., Baan, R., Straif, K., Grosse, Y., Secretan, B., El Ghissassi, F., Benbrahim-Tallaa, L., Guha, N., Freeman, C., Galichet, L., Cogliano, V., 2009. A review of human carcinogens—Part B: biological agents. Lancet Oncol. 10, 321–322.
Cai, P., Gobert, G.N., You, H., McManus, D.P., 2016. The Tao survivorship of schistosomes: implications for schistosomiasis control. Int. J. Parasitol. 46, 453–463.
Cancela, M., Ruétalo, N., Dell'Oca, N., da Silva, E., Smircich, P., Rinaldi, G., Roche, L., Carmona, C., Alvarez-Valín, F., Zaha, A., Tort, J.F., 2010. Survey of transcripts expressed by the invasive juvenile stage of the liver fluke *Fasciola hepatica*. BMC Genom. 11, 227.
Chaiyadet, S., Krueajampa, W., Hipkaeo, W., Plosan, Y., Piratae, S., Sotillo, J., Smout, M., Sripa, B., Brindley, P.J., Loukas, A., Laha, T., 2017. Suppression of mRNAs encoding CD63 family tetraspanins from the carcinogenic liver fluke *Opisthorchis viverrini* results in distinct tegument phenotypes. Sci. Rep. 7, 14342.
Cho, P.Y., Lee, M.J., Kim, T.I., Kang, S.Y., Hong, S.J., 2006. Expressed sequence tag analysis of adult *Clonorchis sinensis*, the Chinese liver fluke. Parasitol. Res. 99, 602–608.
Cho, P.Y., Kim, T.I., Whang, S.M., Hong, S.J., 2008. Gene expression profile of *Clonorchis sinensis* metacercariae. Parasitol. Res. 102, 277–282.
Choi, B.I., Han, J.K., Hong, S.T., Lee, K.H., 2004. Clonorchiasis and cholangiocarcinoma: etiologic relationship and imaging diagnosis. Clin. Microbiol. Rev. 17, 540–552.
Chusorn, P., Namwat, N., Loilome, W., Techasen, A., Pairojkul, C., Khuntikeo, N., Dechakhamphu, A., Talabnin, C., Chan-On, W., Ong, C.K., Teh, B.T., Yongvanit, P., 2013. Overexpression of microRNA-21 regulating PDCD4 during tumorigenesis of liver fluke-associated cholangiocarcinoma contributes to tumor growth and metastasis. Tumour Biol. 34, 1579–1588.
Correnti, J.M., Brindley, P.J., Pearce, E.J., 2005. Long-term suppression of cathepsin B levels by RNA interference retards schistosome growth. Mol. Biochem. Parasitol. 143, 209–215.
Crellen, T., Allan, F., David, S., Durrant, C., Huckvale, T., Holroyd, N., Emery, A.M., Rollinson, D., Aanensen, D.M., Berriman, M., Webster, J.P., Cotton, J.A., 2016. Whole genome resequencing of the human parasite *Schistosoma mansoni* reveals population history and effects of selection. Sci. Rep. 6, 20954.
Cwiklinski, K., Dalton, J.P., Dufresne, P.J., La Course, J., Williams, D.J., Hodgkinson, J., Paterson, S., 2015. The *Fasciola hepatica* genome: gene duplication and polymorphism reveals adaptation to the host environment and the capacity for rapid evolution. Genome Biol. 16, 71.
Dalton, J.P., Brindley, P.J., 1996. Schistosome asparaginyl endopeptidase SM32 in hemoglobin digestion. Parasitol. Today 12, 125.
Fallon, P.G., Doenhoff, M.J., 1994. Drug-resistant schistosomiasis: resistance to praziquantel and oxamniquine induced in *Schistosoma mansoni* in mice is drug specific. Am. J. Trop. Med. Hyg. 51, 83–88.
Folsch, U.R., Wormsley, K.G., 1977. The amino acid composition of rat bile. Experientia 33, 1055–1056.
Fromm, B., Ovchinnikov, V., Høye, E., Bernal, D., Hackenberg, M., Marcilla, A., 2017. On the presence and immunoregulatory functions of extracellular microRNAs in the trematode *Fasciola hepatica*. Parasite Immunol. 39 (2).
Ginsberg, H.N., 1998. Lipoprotein physiology. Endocrinol. Metab. Clin. North Am. 27, 503–519.
Gobert, G.N., McManus, D.P., Nawaratna, S., Moertel, L., Mulvenna, J., Jones, M.K., 2009. Tissue specific profiling of females of *Schistosoma japonicum* by integrated laser microdissection microscopy and microarray analysis. PLoS Negl. Trop. Dis 3, e469.
Gourbal, B.E., Guillou, F., Mitta, G., Sibille, P., Theron, A., Pointier, J.P., Coustau, C., 2008. Excretory-secretory products of larval *Fasciola hepatica* investigated using a two-dimensional proteomic approach. Mol. Biochem. Parasitol. 161, 63–66.

Hagen, J., Young, N.D., Every, A.L., Pagel, C.N., Schnoeller, C., Scheerlinck, J.-P.Y., Gasser, R.B., Kalinna, B.H., 2014. *Omega-1* knockdown in *Schistosoma mansoni* eggs by lentivirus transduction reduces granuloma size *in vivo*. Nat. Commun. 5, 5375.

Hagen, J., Scheerlinck, J.P., Young, N.D., Gasser, R.B., Kalinna, B.H., 2015a. Prospects for vector-based gene silencing to explore immunobiological features of *Schistosoma mansoni*. Adv. Parasitol. 88, 85–122.

Hagen, J., Scheerlinck, J.P., Gasser, R.B., 2015b. Knocking down schistosomes - promise for lentiviral transduction in parasites. Trends Parasitol. 31, 324–332.

Halpern, Z., Rubin, M., Harach, G., Grotto, I., Moser, A., Dvir, A., Lichtenberg, D., 1993. Bile and plasma lipid composition in non-obese normolipidemic subjects with and without cholesterol gallstones. Liver 13, 246–252.

Harinasuta, C., Harinasuta, T., 1984. *Opisthorchis viverrini*: life cycle, intermediate hosts, transmission to man and geographical distribution in Thailand. Arzneimittelforschung 34, 1164–1167.

Harinasuta, T., Riganti, M., Bunnag, D., 1984. *Opisthorchis viverrini* infection: pathogenesis and clinical features. Arzneimittelforschung 34, 1167–1169.

Harris, T.W., Baran, J., Bieri, T., Cabunoc, A., Chan, J., Chen, W.J., Davis, P., Done, J., Grove, C., Howe, K., Kishore, R., Lee, R., Li, Y., Muller, H.M., Nakamura, C., Ozersky, P., Paulini, M., Raciti, D., Schindelman, G., Tuli, M.A., Van Auken, K., Wang, D., Wang, X., Williams, G., Wong, J.D., Yook, K., Schedl, T., Hodgkin, J., Berriman, M., Kersey, P., Spieth, J., Stein, L., Sternberg, P.W., 2014. WormBase 2014: new views of curated biology. Nucleic Acids Res. 42, D789eD793.

Haswell-Elkins, M.R., Mairiang, E., Mairiang, P., Chaiyakum, J., Chamadol, N., Loapaiboon, V., Sithithaworn, P., Elkins, D.B., 1994. Cross-sectional study of *Opisthorchis viverrini* infection and cholangiocarcinoma in communities within a high-risk area in northeast Thailand. Int. J. Cancer 59, 505–509.

Honjo, S., Srivatanakul, P., Sriplung, H., Kikukawa, H., Hanai, S., Uchida, K., Todoroki, T., Jedpiyawongse, A., Kittiwatanachot, P., Sripa, B., Deerasamee, S., Miwa, M., 2005. Genetic and environmental determinants of risk for cholangiocarcinoma via *Opisthorchis viverrini* in a densely infested area in Nakhon Phanom, northeast Thailand. Int. J. Cancer 117, 854–860.

Hou, P.C., 1955. The pathology of *Clonorchis sinensis* infestation of the liver. J. Pathol. Bacteriol. 70, 53–64.

Hu, F., Hu, X., Ma, C., Zhao, J., Xu, J., Yu, X., 2009. Molecular characterization of a novel *Clonorchis sinensis* secretory phospholipase A2 and investigation of its potential contribution to hepatic fibrosis. Mol. Biochem. Parasitol. 167, 127–134.

Huang, Y., Chen, W., Wang, X., Liu, H., Chen, Y., Guo, L., Luo, F., Sun, J., Mao, Q., Liang, P., Xie, Z., Zhou, C., Tian, Y., Lv, X., Huang, L., Zhou, J., Hu, Y., Li, R., Zhang, F., Lei, H., Li, W., Hu, X., Liang, C., Xu, J., Li, X., Yu, X., 2013. The carcinogenic liver fluke, *Clonorchis sinensis*: new assembly, reannotation and analysis of the genome and characterization of tissue transcriptomes. PLoS One 8, e54732.

Inohara, N., Nunez, G., 2002. ML - a conserved domain involved in innate immunity and lipid metabolism. Trends Biochem. Sci. 27, 219–221.

Jex, A.R., Young, N.D., Sripa, J., Hall, R.S., Scheerlinck, J.P., Laha, T., Sripa, B., Gasser, R.B., 2012. Molecular changes in *Opisthorchis viverrini* (Southeast Asian Liver Fluke) during the transition from the juvenile to the adult stage. PLoS Negl. Trop. Dis 6, e1916.

Ju, J.W., Joo, H.N., Lee, M.R., Cho, S.H., Cheun, H.I., Kim, J.Y., Lee, Y.H., Lee, K.J., Sohn, W.M., Kim, D.M., Kim, I.C., Park, B.C., Kim, T.S., 2009. Identification of a serodiagnostic antigen, legumain, by immunoproteomic analysis of excretory-secretory products of *Clonorchis sinensis* adult worms. Proteomics 9, 3066–3078.

Kaewkong, W., Choochote, W., Kanla, P., Maleewong, W., Intapan, P.M., Wongkham, S., Wongkham, C., 2012. Chromosomes and karyotype analysis of a liver fluke, *Opisthorchis viverrini*, by scanning electron microscopy. Parasitol. Int. 61, 504–507.
Kang, J.M., Bahk, Y.Y., Cho, P.Y., Hong, S.J., Kim, T.S., Sohn, W.M., Na, B.K., 2010. A family of cathepsin F cysteine proteases of *Clonorchis sinensis* is the major secreted proteins that are expressed in the intestine of the parasite. Mol. Biochem. Parasitol. 170, 7–16.
Keiser, J., Shu-Hua, X., Jian, X., Zhen-San, C., Odermatt, P., Tesana, S., Tanner, M., Utzinger, J., 2006. Effect of artesunate and artemether against *Clonorchis sinensis* and *Opisthorchis viverrini* in rodent models. Int. J. Antimicrob. Agents 28, 370–373.
Keiser, J., Utzinger, J., 2009. Food-borne trematodiases. Clin. Microbiol. Rev. 22, 466–483.
Keiser, J., Xiao, S.H., Smith, T.A., Utzinger, J., 2009. Combination chemotherapy against *Clonorchis sinensis*: experiments with artemether, artesunate, OZ78, praziquantel, and tribendimidine in a rat model. Antimicrob. Agents Chemother. 53, 3770–3776.
Keiser, J., Adelfio, R., Vargas, M., Odermatt, P., Tesana, S., 2013. Activity of tribendimidine and praziquantel combination therapy against the liver fluke *Opisthorchis viverrini* in vitro and in vivo. J. Helminthol. 87, 252–256.
Kiger, L., Rashid, A.K., Griffon, N., Haque, M., Moens, L., Gibson, Q.H., Poyart, C., Marden, M.C., 1998. Trematode hemoglobins show exceptionally high oxygen affinity. Biophys. J. 75, 990–998.
Kim, E.M., Kim, J.S., Choi, M.H., Hong, S.T., Bae, Y.M., 2008. Effects of excretory/secretory products from *Clonorchis sinensis* and the carcinogen dimethylnitrosamine on the proliferation and cell cycle modulation of human epithelial HEK293T cells. Korean J. Parasitol. 46, 127–132.
Korhonen, P.K., Young, N.D., Gasser, R.B., 2016. Making sense of genomes of parasitic worms: tackling bioinformatic challenges. Biotechnol. Adv. 34, 663–686.
Krautz-Peterson, G., Simoes, M., Faghiri, Z., Ndegwa, D., Oliveira, G., Shoemaker, C.B., Skelly, P.J., 2010. Suppressing glucose transporter gene expression in schistosomes impairs parasite feeding and decreases survival in the mammalian host. PLoS Pathog. 6, e1000932.
Kurathong, S., Lerdverasirikul, P., Wongpaitoon, V., Pramoolsinsap, C., Upatham, E.S., 1987. *Opisthorchis viverrini* infection in rural and urban communities in northeast Thailand. Trans. R. Soc. Trop. Med. Hyg. 1987 (81), 411–444.
Laha, T., Pinlaor, P., Mulvenna, J., Sripa, B., Sripa, M., Smout, M.J., Gasser, R.B., Brindley, P.J., Loukas, A., 2007. Gene discovery for the carcinogenic human liver fluke, *Opisthorchis viverrini*. BMC Genom. 8, 189.
Lechner, S., Müller-Ladner, U., Schlottmann, K., Jung, B., McClelland, M., Rüschoff, J., Welsh, J., Schölmerich, J., Kullmann, F., 2002. Bile acids mimic oxidative stress induced upregulation of thioredoxin reductase in colon cancer cell lines. Carcinogenesis 23, 1281–1288.
Lee, J.H., Rim, H.J., Bak, U.B., 1993. Effect of *Clonorchis sinensis* infection and dimethylnitrosamine administration on the induction of cholangiocarcinoma in Syrian golden hamsters. Korean J. Parasitol. 31, 21–30.
Lim, M.K., Ju, Y.H., Franceschi, S., Oh, J.K., Kong, H.J., Hwang, S.S., Park, S.K., Cho, S.I., Sohn, W.M., Kim, D.I., Yoo, K.Y., Hong, S.T., Shin, H.R., 2006. *Clonorchis sinensis* infection and increasing risk of cholangiocarcinoma in the Republic of Korea. Am. J. Trop. Med. Hyg. 75, 93–96.
Madoui, M.A., Engelen, S., Cruaud, C., Belser, C., Bertrand, L., Alberti, A., Lemainque, A., Wincker, P., Aury, J.M., 2015. Genome assembly using nanopore-guided long and error-free DNA reads. BMC Genom. 16, 327.
Mairiang, E., Elkins, D.B., Mairiang, P., Chaiyakum, J., Chamadol, N., Loapaiboon, V., Posri, S., Sithithaworn, P., Haswell-Elkins, M., 1992. Relationship between intensity of *Opisthorchis viverrini* infection and hepatobiliary disease detected by ultrasonography. J. Gastroenterol. Hepatol. 7, 17–21.

Masyuk, A.I., Masyuk, T.V., Tietz, P.S., Splinter, P.L., LaRusso, N.F., 2002. Intrahepatic bile ducts transport water in response to absorbed glucose. Am. J. Physiol. Cell Physiol. 283, C785—C791.

McSorley, H.J., Hewitson, J.P., Maizels, R.M., 2013. Immunomodulation by helminth parasites: defining mechanisms and mediators. Int. J. Parasitol. 43, 301—310.

Min, H.K., 1984. *Clonorchis sinensis*: pathogenesis and clinical features of infection. Arzneimittelforschung 34, 1151—1153.

Mulvenna, J., Sripa, B., Brindley, P.J., Gorman, J., Jones, M.K., Colgrave, M.L., Jones, A., Nawaratna, S., Laha, T., Suttiprapa, S., Smout, M.J., Loukas, A., 2010. The secreted and surface proteomes of the adult stage of the carcinogenic human liver fluke *Opisthorchis viverrini*. Proteomics 10, 1063—1078.

Nagaraj, S.H., Gasser, R.B., Ranganathan, S., 2007. A hitchhiker's guide to expressed sequence tag (EST) analysis. Brief. Bioinform 8, 6—21.

Nair, S.S., Bommana, A., Bethony, J.M., Lyon, A.J., Ohshiro, K., Pakala, S.B., Rinaldi, G., Keegan, B., Suttiprapa, S., Periago, M.V., Hotez, P.J., Brindley, P.J., Kumar, R., 2011. The metastasis-associated protein-1 gene encodes a host permissive factor for schistosomiasis, a leading global cause of inflammation and cancer. Hepatology 54, 285—295.

Nawaratna, S.S., McManus, D.P., Moertel, L., Gobert, G.N., Jones, M.K., 2011. Gene atlasing of digestive and reproductive tissues in *Schistosoma mansoni*. PLoS Negl. Trop. Dis. 5 (4), e1043.

Nawaratna, S.S., Gobert, G.N., Willis, C., Chuah, C., McManus, D.P., Jones, M.K., 2014. Transcriptional profiling of the oesophageal gland region of male worms of *Schistosoma mansoni*. Mol. Biochem. Parasitol. 196, 82—89.

Nawrocki, E.P., Burge, S.W., Bateman, A., Daub, J., Eberhardt, R.Y., Eddy, S.R., Floden, E.W., Gardner, P.P., Jones, T.A., Tate, J., Finn, R.D., 2015. Rfam 12.0: updates to the RNA families database. Nucleic Acids Res. 43, D130—D137.

Ong, C.K., Subimerb, C., Pairojkul, C., Wongkham, S., Cutcutache, I., Yu, W., McPherson, J.R., Allen, G.E., Ng, C.C., Wong, B.H., Myint, S.S., Rajasegaran, V., Heng, H.L., Gan, A., Zang, Z.J., Wu, Y., Wu, J., Lee, M.H., Huang, D., Ong, P., Chan-on, W., Cao, Y., Qian, C.N., Lim, K.H., Ooi, A., Dykema, K., Furge, K., Kukongviriyapan, V., Sripa, B., Wongkham, C., Yongvanit, P., Futreal, P.A., Bhudhisawasdi, V., Rozen, S., Tan, P., Teh, B.T., 2012. Exome sequencing of liver fluke-associated cholangiocarcinoma. Nat. Genet. 44, 690—693.

Orido, Y., 1990. Development of the excretory bladder of the lung fluke *Paragonimus ohirai* (Trematoda: Troglotrematidae). J. Parasitol. 76, 205—211.

Ovchinnikov, V.Y., Mordvinov, V.A., Fromm, B., 2017. Extreme conservation of miRNA complements in opisthorchiids. Parasitol. Int. 66, 773—776.

Papatpremsiri, A., Smout, M.J., Loukas, A., Brindley, P.J., Sripa, B., Laha, T., 2015. Suppression of *Ov-grn-1* encoding granulin of *Opisthorchis viverrini* inhibits proliferation of biliary epithelial cells. Exp. Parasitol. 148, 17—23.

Park, G.M., Im, K., Huh, S., Yong, T.S., 2000. Chromosomes of the liver fluke, *Clonorchis sinensis*. Korean J. Parasitol. 38, 201—206.

Peng, J., Feng, Y., Rinaldi, G., Yonglitthipagon, P., Easley, S.E., Laha, T., Pairojkul, C., Bhudhisawasdi, V., Sripa, B., Brindley, P.J., Mulvenna, J.P., Bethony, J.M., Plieskatt, J.L., 2014. The miRNAome of *Opisthorchis viverrini* induced intrahepatic cholangiocarcinoma. Genom. Data 2, 274—279.

Petney, T.N., Andrews, R.H., Saijuntha, W., Wenz-Mücke, A., Sithithaworn, P., 2013. Int. J. Parasitol. 43, 1031—1046.

Plieskatt, J.L., Rinaldi, G., Feng, Y., Peng, J., Yonglitthipagon, P., Easley, S., Laha, T., Pairojkul, C., Bhudhisawasdi, V., Sripa, B., Brindley, P.J., Mulvenna, J.P., Bethony, J.M., 2014. Distinct miRNA signatures associate with subtypes of cholangiocarcinoma from infection with the tumourigenic liver fluke *Opisthorchis viverrini*. J. Hepatol. 61, 850—858.

Plieskatt, J.L., Rinaldi, G., Feng, Y., Peng, J., Easley, S., Jia, X., Potriquet, J., Pairojkul, C., Bhudhisawasdi, V., Sripa, B., Brindley, P.J., Bethony, J., Mulvenna, J., 2015. A microRNA profile associated with *Opisthorchis viverrini*-induced cholangiocarcinoma in tissue and plasma. BMC Canc. 15, 309.
Robinson, M.W., Donnelly, S., Hutchinson, A.T., To, J., Taylor, N.L., Norton, R.S., Perugini, M.A., Dalton, J.P., 2011. A family of helminth molecules that modulate innate cell responses via molecular mimicry of host antimicrobial peptides. PLoS Pathog. 7, e1002042.
Safonova, Y., Bankevich, A., Pevzner, P.A., 2015. dipSPAdes: assembler for highly polymorphic diploid genomes. J. Comput. Biol. 22, 528e545.
Genome Sequencing and Functional Analysis Consortium, 2009. The *Schistosoma japonicum* genome reveals features of host-parasite interplay. Nature 460, 345−351.
Sharma, R., Yang, Y., Sharma, A., Awasthi, S., Awasthi, Y.C., 2004. Antioxidant role of glutathione S-transferases: protection against oxidant toxicity and regulation of stress-mediated apoptosis. Antioxid. Redox Signal. 6, 289−300.
Sharon, I., Kertesz, M., Hug, L.A., Pushkarev, D., Blauwkamp, T.A., Castelle, C.J., Amirebrahimi, M., Thomas, B.C., Burstein, D., Tringe, S.G., Williams, K.H., Banfield, J.F., 2015. Accurate, multi-kb reads resolve complex populations and detect rare microorganisms. Genome Res. 25, 534e543.
Shin, H.R., Oh, J.K., Masuyer, E., Curado, M.P., Bouvard, V., Fang, Y.Y., Wiangnon, S., Sripa, B., Hong, S.T., 2010. Epidemiology of cholangiocarcinoma: an update focusing on risk factors. Cancer Sci. 101, 579−585.
Short, R.B., Menzel, M.Y., 1960. Chromosomes of nine species of schistosomes. J. Parasitol. 46, 273−287.
Silakit, R., Loilome, W., Yongvanit, P., Chusorn, P., Techasen, A., Boonmars, T., Khuntikeo, N., Chamadol, N., Pairojkul, C., Namwat, N., 2014. Circulating miR-192 in liver fluke-associated cholangiocarcinoma patients: a prospective prognostic indicator. J. Hepatobiliary Pancreat. Sci. 21, 864−872.
Silakit, R., Loilome, W., Yongvanit, P., Thongchot, S., Sithithaworn, P., Boonmars, T., Koonmee, S., Titapun, A., Khuntikeo, N., Chamadol, N., Techasen, A., Namwat, N., 2017. Urinary microRNA-192 and microRNA-21 as potential indicators for liver fluke-associated cholangiocarcinoma risk group. Parasitol. Int. 66, 479−485.
Sithithaworn, P., Haswell-Elkins, M., 2003. Epidemiology of *Opisthorchis viverrini*. Acta Trop. 88, 187−194.
Sithithaworn, P., Haswell-Elkins, M.R., Mairiang, P., Satarug, S., Mairiang, E., Vatanasapt, V., Elkins, D.B., 1994. Parasite-associated morbidity: liver fluke infection and bile duct cancer in northeast Thailand. Int. J. Parasitol. 24, 833−843.
Smith, A.M., Dowd, A.J., Heffernan, M., Robertson, C.D., Dalton, J.P., 1993. *Fasciola hepatica*: a secreted cathepsin L-like proteinase cleaves host immunoglobulin. Int. J. Parasitol. 23, 977−983.
Smout, M.J., Laha, T., Mulvenna, J., Sripa, B., Suttiprapa, S., Jones, A., Brindley, P.J., Loukas, A., 2009. A granulin-like growth factor secreted by the carcinogenic liver fluke, *Opisthorchis viverrini*, promotes proliferation of host cells. PLoS Pathog. 5, e1000611.
Spencer, W.C., McWhirter, R., Miller, T., Strasbourger, P., Thompson, O., Hillier, L.W., Waterston, R.H., Miller 3rd, D.M., 2014. Isolation of specific neurons from *C. elegans* larvae for gene expression profiling. PLoS One 9, e112102.
Spitz, F., Furlong, E.E., 2012. Transcription factors: from enhancer binding to developmental control. Nat. Rev. Genet. 13, 613−626.
Sripa, B., Pairojkul, C., 2008. Cholangiocarcinoma: lessons from Thailand. Curr. Opin. Gastroenterol. 24, 349−356.

Sripa, B., Kaewkes, S., Sithithaworn, P., Mairiang, E., Laha, T., Smout, M.J., Pairojkul, C., Bhudhisawasdi, V., Tesana, S., Thinkamrop, B., Bethony, J.M., Loukas, A., Brindley, P.J., 2007. Liver fluke induces cholangiocarcinoma. PLoS Med. 4, e201.

Sripa, B., Mairiang, E., Thinkhamrop, B., Laha, T., Kaewkes, S., Sithithaworn, P., Tessana, S., Loukas, A., Brindley, P.J., Bethony, J.M., 2009. Advanced periductal fibrosis from infection with the carcinogenic human liver fluke *Opisthorchis viverrini* correlates with elevated levels of interleukin-6. Hepatology 50, 1273–1281.

Sripa, J., Pinlaor, P., Brindley, P.J., Sripa, B., Kaewkes, S., Robinson, M.W., Young, N.D., Gasser, R.B., Loukas, A., Laha, T., 2011. RNA interference targeting cathepsin B of the carcinogenic liver fluke, *Opisthorchis viverrini*. Parasitol. Int. 60, 283–288.

Sripa, B., Brindley, P.J., Mulvenna, J., Laha, T., Smout, M.J., Mairiang, E., Bethony, J.M., Loukas, A., 2012. The tumorigenic liver fluke *Opisthorchis viverrini*—multiple pathways to cancer. Trends Parasitol. 28, 395–407.

Sripa, B., Tangkawattana, S., Laha, T., Kaewkes, S., Mallory, F.F., Smith, J.F., Wilcox, B.A., 2015. Toward integrated opisthorchiasis control in northeast Thailand: the Lawa project. Acta Trop. 141, 361–367.

Subramanian, K., Balch, W.E., 2008. NPC1/NPC2 function as a tag team duo to mobilize cholesterol. Proc. Natl Acad. Sci. USA 105, 15223–15224.

Thamavit, W., Bhamarapravati, N., Sahaphong, S., Vajrasthira, S., Angsubhakorn, S., 1978. Effects of dimethylnitrosamine on induction of cholangiocarcinoma in *Opisthorchis viverrini*- infected Syrian golden hamsters. Cancer Res. 38, 4634–4639.

Thuwajit, C., Thuwajit, P., Uchida, K., Daorueang, D., Kaewkes, S., Wongkham, S., Miwa, M., 2006. Gene expression profiling defined pathways correlated with fibroblast cell proliferation induced by *Opisthorchis viverrini* excretory/secretory product. World J. Gastroenterol. 12, 3585–3592.

Trauner, M., Boyer, J.L., 2003. Bile salt transporters: molecular characterization, function, and regulation. Physiol. Rev. 83, 633–671.

Tsai, I.J., Zarowiecki, M., Holroyd, N., Garciarrubio, A., Sánchez-Flores, A., Brooks, K.L., Tracey, A., Bobes, R.J., Fragoso, G., Sciutto, E., Aslett, M., Beasley, H., Bennett, H.M., Cai, X., Camicia, F., Clark, R., Cucher, M., De Silva, N., Day, T.A., Deplazes, P., Estrada, K., Fernández, C., Holland, P.W.H., Hou, J., Hu, S., Huckvale, T., Hung, S.S., Kamenetzky, L., Keane, J.A., Kiss, F., Koziol, U., Lambert, O., Liu, K., Luo, X., Luo, Y., Macchiaroli, N., Nichol, S., Paps, J., Parkinson, J., Pouchkina-Stantcheva, N., Riddiford, N., Rosenzvit, M., Salinas, G., Wasmuth, J.D., Zamanian, M., Zheng, Y., Genome Consortium, Cai, J., Soberón, X., Olson, P.D., Laclette, J.P., Brehm, K., Berriman, M., 2013. The genomes of four tapeworm species reveal adaptations to parasitism. Nature 496, 57–63.

van Hellemond, J.J., van der Klei, A., van Weelden, S.W., Tielens, A.G., 2003. Biochemical and evolutionary aspects of anaerobically functioning mitochondria. Philos. Trans. R. Soc. Lond. B Biol. Sci. 358, 205–213 discussion 213–215.

Wang, D., Korhonen, P.K., Gasser, R.B., Young, N.D., 2018. Improved genomic resources and new bioinformatic workflow for the carcinogenic parasite *Clonorchis sinensis*: biotechnological implications. Biotechnol. Adv. (Submitted Jan. 2018).

Wongratanacheewin, S., Sermswan, R.W., Sirisinha, S., 2003. Immunology and molecular biology of *Opisthorchis viverrini* infection. Acta Trop. 88, 195–207.

Young, N.D., Campbell, B.E., Hall, R.S., Jex, A.R., Cantacessi, C., Laha, T., Sohn, W.M., Sripa, B., Loukas, A., Brindley, P.J., Gasser, R.B., 2010a. Unlocking the transcriptomes of two carcinogenic parasites, *Clonorchis sinensis* and *Opisthorchis viverrini*. PLoS Negl. Trop. Dis. 4, e719.

Young, N.D., Hall, R.S., Jex, A.J., Cantacessi, C., Gasser, R.B., 2010b. Elucidating the transcriptome of *Fasciola hepatica* — a key to fundamental and biotechnological discoveries for a neglected parasite. Biotechnol. Adv. 28, 222–231.

Young, N.D., Jex, A.R., Li, B., Liu, S., Yang, L., Xiong, Z., Li, Y., Cantacessi, C., Hall, R.S., Xu, X., Chen, F., Wu, X., Zerlotini, A., Oliveira, G., Hofmann, A., Zhang, G., Fang, X., Kang, Y., Campbell, B.E., Loukas, A., Ranganathan, S., Rollinson, D., Rinaldi, G., Brindley, P.J., Yang, H., Wang, J., Wang, Gasser, R.B., 2012. Whole-genome sequence of *Schistosoma haematobium*. Nature Genet. 44, 221–225.

Young, N.D., Nagarajan, N., Lin, S.J., Korhonen, P.K., Jex, A.R., Hall, R.S., Safavi-Hemami, H., Kaewkong, W., Bertrand, D., Gao, S., Seet, Q., Wongkham, S., Teh, B.T., Wongkham, C., Intapan, P.M., Maleewong, W., Yang, X., Hu, M., Wang, Z., Hofmann, A., Sternberg, P.W., Tan, P., Wang, J., Gasser, R.B., 2014. The *Opisthorchis viverrini* genome provides insights into life in the bile duct. Nat. Commun. Now. 5, 4378.

Young, N.D., Chan, K.G., Korhonen, P.K., Min Chong, T., Ee, R., Mohandas, N., Koehler, A.V., Lim, Y.L., Hofmann, A., Jex, A.R., Qian, B., Chilton, N.B., Gobert, G.N., McManus, D.P., Tan, P., Webster, B.L., Rollinson, D., Gasser, R.B., 2015. Exploring molecular variation in *Schistosoma japonicum* in China. Sci. Rep. 5, 17345.

Zdobnov, E.M., Apweiler, R., 2001. InterProScan — an integration platform for the signature- recognition methods in InterPro. Bioinformatics 17, 847–848.

CHAPTER SIX

Opisthorchiasis-Induced Cholangiocarcinoma: How Innate Immunity May Cause Cancer

Steven W. Edwards*,[1], Edward M. Spofford*, Charlotte Price*, Helen L. Wright*,[a], Kanin Salao[§], Sutas Suttiprapa[¶,||] and Banchob Sripa[¶,||]

*Institute of Integrative Biology, University of Liverpool, Liverpool, United Kingdom
[§]Department of Microbiology, Faculty of Medicine, Khon Kaen University, Khon Kaen, Thailand
[¶]Tropical Medicine Graduate Program, Faculty of Medicine, Khon Kaen University, Khon Kaen, Thailand
[||]WHO Collaborating Centre for Research and Control of Opisthorchiasis (Southeast Asian Liver Fluke Disease), Department of Pathology, Faculty of Medicine, Khon Kaen University, Khon Kaen, Thailand
[1]Corresponding author: Email: s.w.edwards@liverpool.ac.uk

Contents

[a] Current address: Institute of Ageing and Chronic Disease, University of Liverpool, Liverpool, United Kingdom

Advances in Parasitology, Volume 101
ISSN 0065-308X
https://doi.org/10.1016/bs.apar.2018.05.006

Abstract

Innate, inflammatory responses towards persistent *Opisthorchis viverrini* (OV) infection are likely to contribute to the development of cholangiocarcinoma (CCA), a liver cancer that is rare in the West but prevalent in Greater Mekong Subregion countries in Southeast Asia. Infection results in the infiltration of innate immune cells into the bile ducts and subsequent activation of inflammatory immune responses that fail to clear OV but instead may damage local tissues within the bile ducts. Not all patients infected with OV develop CCA, and so tumourigenesis may be dependent on multiple factors including the magnitude of the inflammatory response that is activated in infected individuals. The purpose of this review is to summarize how innate immune responses may promote tumourigenesis following OV infection and if such responses can be used to predict CCA onset in OV-infected individuals. It also hypothesizes on the role that *Helicobacterspp.*, which are associated with liver fluke infections, may play in activation of the innate the immune system to promote tissue damage and persistent inflammation leading to CCA.

1. INTRODUCTION

1.1 *Opisthorchis viverrini*, An Endemic Disease Throughout Southeast Asia

Infection with the liver fluke *Opisthorchis viverrini* (OV), or opisthorchiasis, has long been linked with the onset of cholangiocarcinoma (CCA). The parasite infects human hosts (and other piscivorous animals) when they eat OV-infected, undercooked/raw cyprinoid fish. Rates of OV infection are especially high in the northeastern region of Thailand (Sithithaworn et al., 2012a; Yeoh et al., 2015), where persistent infection is a consequence of eating raw fish, often as part of traditional dishes such as '*Koi Pla*' and '*Pla Som*'. Numerous control programmes have been implemented in OV endemic areas (Sripa et al., 2015), but despite initial decreases in opisthorchiasis following the introduction of such programmes, rates of CCA remain high in Thailand and neighbouring countries, such as Lao-PDR, Cambodia and Vietnam. This disease therefore continues to place a severe burden on both the health services and the regional economy (Andrews et al., 2008), highlighting an urgent need for an improved understanding of OV pathogenesis and of the processes that lead to the development of CCA. OV infection and liver disease are often asymptomatic, and there is currently no cure for CCA, and often metastasis has occurred before detection, with patients surviving for only short periods (sometimes less than 4 months) post diagnosis (Luvira et al., 2016). To date, the antihelminthic drug praziquantel can be used as a chemotherapeutic, but this can only treat OV infection, and not CCA itself.

There is considerable heterogeneity in pathological outcomes following OV infection, with some individuals infected for their entire lives without showing any clinical symptoms. However, approximately 25% of those infected with OV develop advanced periductal fibrosis and approximately 1% develop CCA (Mairiang et al., 2012). The reasons for this disease heterogeneity are not fully understood, but it has been hypothesized that a 'proinflammatory phenotype' predisposes particular individuals to develop advanced periductal fibrosis (APF) following OV infection that may lead to CCA (Sripa et al., 2012a). If it is the case that OV-induced CCA is related to patient-specific immune responses, then identifying these responses may make the onset of cancer more predictable or perhaps preventable. To date, whilst many reviews have suggested a possible link between OV-induced CCA and adaptive inflammatory responses, there is a paucity of literature describing specific innate immune cell responses to OV. If the nature of the nonspecific, innate immune responses towards OV can be identified and are confirmed to be responsible for causing or contributing to the development of CCA, these may represent novel targets for treatment and prevention of OV-induced CCA (Sripa et al., 2012a). In addition, these responses may serve as biomarkers that could identify individuals at risk of developing CCA following OV infection and hence enable a more targeted approach to prevention and therapy. This review aims to improve our understanding of the innate immune responses elicited by OV infection and discusses how this may be linked to the onset of CCA. It highlights a possible link between liver fluke infection, *Helicobacter* coinfection and innate immune cell activation in disease pathology.

1.2 *Opisthorchis viverrini*: An Infectious Agent Capable of Driving Cancer

Infection with OV can cause jaundice, cholangitis and CCA, which is a malignancy of the bile duct epithelium, and relatively rare worldwide but highly prevalent in Mekong countries (Sithithaworn et al., 2012b). CCA is usually associated with OV infection, but alcohol, smoking and diets containing nitrogenous compounds may all contribute towards malignancy (Sithithaworn et al., 2014). There are two recognized types of CCA, OV-associated and non-OV-associated, which have differential carcinoma expression profiles, but the 5-year survival rate for both pathologies is <5% (Sripa and Pairojkul, 2008; Ghouri et al., 2015). The initiation of CCA in opisthorchiasis patients occurs in areas of inflammation, fibrosis and proliferating epithelial cells that may have an increased risk of tumourigenesis when also exposed to carcinogens (Sripa et al., 2007). Due to its association with CCA, OV is recognized as a Group 1 carcinogen and one of few infections that have known carcinogenic properties; others include Hepatitis B and *Helicobacter pylori* infections (IARC, 2011). Other liver flukes endemic in Southeast Asia, including *Clonorchis sinensis* and *Opisthorchis felineus* (Lim, 2011), are less frequently associated with CCA.

There are three hypothesized causes of OV-induced CCA: (1) chronic or persistent inflammatory responses towards persistent opisthorchiasis; (2) anti-apoptotic/immunomodulatory properties of *O. viverrini* excretory/secretory (OVES) products products; and (3) structural damage caused by OV feeding on epithelial cells (Sripa et al., 2007; Sripa et al., 2012a). All three processes may be involved and may contribute to CCA. Inflammatory responses, particularly those that result in the generation of free radicals, may inflict oxidative and nitrogenous damage on the DNA of local cells, a well-established hallmark of cancer (Murata et al., 2012; Yongvanit et al., 2012a). The recent discovery of an association between OV infection with *Helicobacter* infections (*H. pylori* and *H. bilis*) (Deenonpoe et al., 2015, 2017) and the known effects of these bacteria in both cancer pathogenesis and immune cell activation suggest an additional mechanism to explain how inflammation may contribute to the development of CCA (Sripa et al., 2017).

The fact that opisthorchiasis is in decline in some regions, but CCA remains prevalent, suggests that either OV infection alone cannot directly induce CCA, and that a multitude of factors are involved in CCA pathogenesis, or alternatively, that OV infection is an initiator of cancer. Thus, if tissue damage is initiated by OV-induced inflammation, then a vicious cycle of inflammation—tissue damage: inflammation leading to CCA—may exist even when the OV infection has been cleared by treatment (Fig. 1). This concept is supported in hamster models, where CCA only develops when OV infection is combined with exposure to subcarcinogenic dose of nitrosamines such as N-nitrosodimethylamine (NDMA) (Thamavit et al., 1978; Boonmars et al., 2011). Nitrosamines from fermented fish are common in the diet of those who eat traditional, raw fish meals and are also linked with carcinogenesis (Bartsch and Montesano, 1984). Infection with OV alone causes hamsters to develop intrahepatic and extrahepatic biliary damage/abnormalities associated with CCA (Thamavit et al., 1978), but not the onset of cancer itself. It would appear that infected hosts cannot produce an

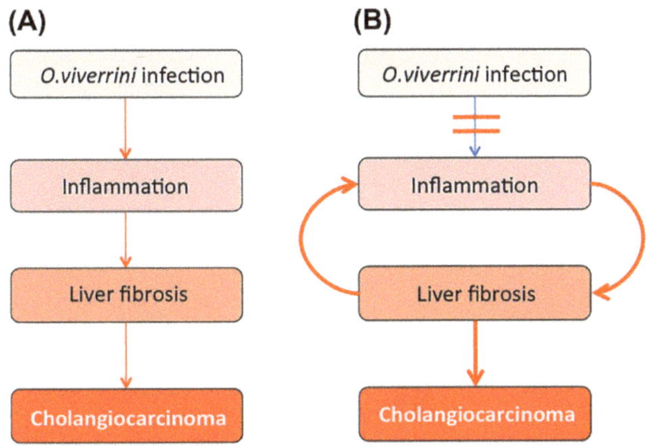

Figure 1 Proposed role of persistent inflammation in the development of cholangiocarcinoma following infection with the liver fluke, *Opisthorchis viverrini* (*O. viverrini*). In (A), infection with the liver fluke leads to inflammation that can (along with other factors, see Fig. 2) result in the development of fibrosis and subsequently cholangiocarcinoma. In (B), it is proposed that, in some individuals, inflammation can persist even after the parasite has been cleared (e.g., by praziquantel treatment). This persistent inflammation may lead to a vicious cycle of inflammation/damage that can lead to fibrosis and ultimately cholangiocarcinoma. In this proposed model, inflammation can drive liver fibrosis, whereas liver fibrosis can also drive further cycles of inflammation.

efficient immune response to clear the parasite and/or that OV is capable of evading or modulating immune responses, which allows its survival (McSorley et al., 2013). The persistent inflammatory responses directed towards OV indirectly promote tumourigenesis by providing the suitable microenvironment for tumours to develop but may require additional carcinogens such as nitrosamines and alcohol, to initiate tumourigenesis (Sripa et al., 2007, Fig. 2). It is feasible that OV infection physically blocks the bile ducts, with the result that the accumulated endogenous nitrosamines and free radicals have prolonged exposure to local cells and attain very high local concentrations, increasing the chances of DNA damage (Sripa et al., 2007).

The hypothesized 'proinflammatory phenotype' may only reveal itself in instances of long-term or repeated infections, such as during OV infection or untreated *Helicobacter* infections, and this unresolved and sustained inflammation may lead to tissue damage and initiation of carcinogenesis and CCA (Sripa et al., 2012a, 2017). There have been several reviews that discuss the link between inflammatory responses and cancer, which may also be applicable to OV-induced CCA (Grivennikov et al., 2010; Balkwill and

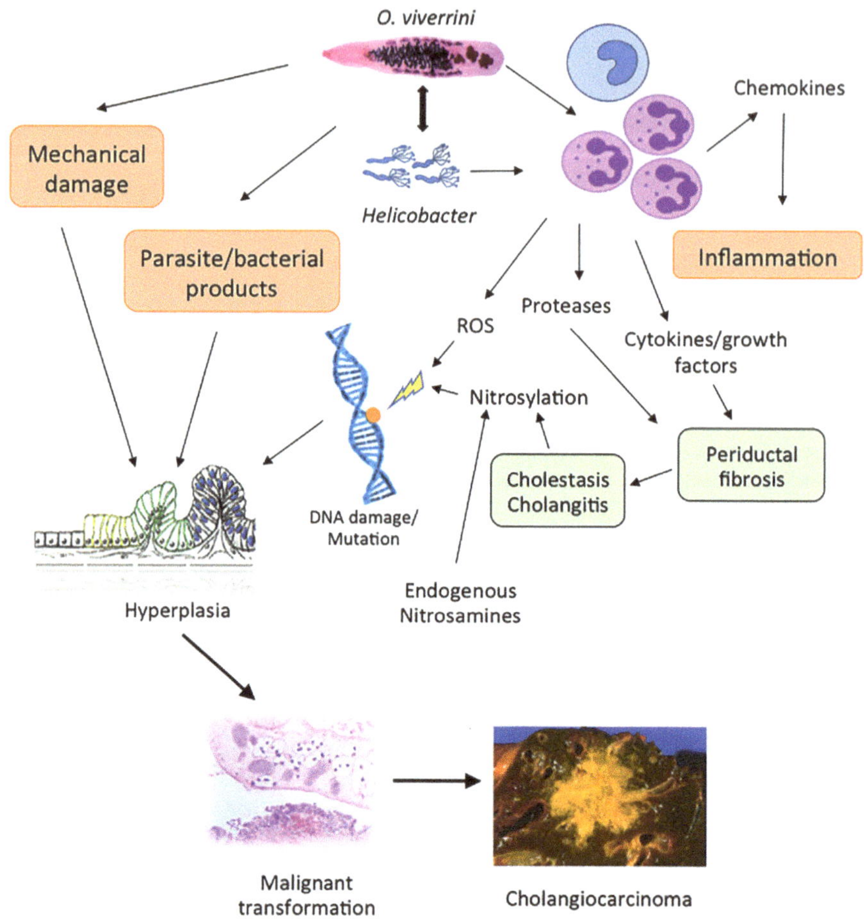

Figure 2 Proposed interplay between parasites, *Helicobacter* and inflammation in the events that lead to periductal fibrosis, DNA mutations, hyperplasia and malignant transformations that can lead to the development of cholangiocarcinoma. *Modified from Sripa, B., Deenonpoe, R., Brindley, P.J., 2017. Coinfections with liver fluke and* Helicobacter *species: a paradigm change in pathogenesis of opisthorchiasis and cholangiocarcinoma? Parasitol. Int. 66, 383–389.*

Mantovani, 2001; Lopez-Novoa and Nieto, 2009). Parallels may also be drawn with *H. pylori* infection, where almost 50% of the world's population are infected, yet only some develop the associated gastric cancer. Opisthorchiasis therefore represents another example of an inflammatory disease that elevates the possibility of cancer and may have some parallels with diseases such as rheumatoid arthritis and atherosclerosis (Chen et al., 2011; Pittet and Swirski, 2011).

2. IMMUNE ACTIVATION

2.1 Regulation of Immune Cell Activation During Helminth Infections

Inflammation in the context of helminth infection differs from other classical acute, inflammatory responses because helminths possess immunomodulatory properties. Helminths are capable of regulating the normal T helper 2 (Th2) responses against pathogens, mediated by immunoglobulin (Ig)E, eosinophils, mast cells and Th2 cytokines such as interleukin (IL)-4 and IL-5 (Maizels and Yazdanbakhsh, 2003; Maizels et al., 2014). Unlike microbial pathogen recognition, there are no clearly identified pathogen-associated molecular patterns on helminths that facilitate their recognition by pattern recognition receptors (PRRs) on immune cells (Perrigoue et al., 2008). Helminth-mediated immunomodulation is typically associated with increased antiinflammatory mediators such as IL-10 levels and Treg activity, a finding validated in mouse models (Wilson et al., 2005). Helminths modulate host immunity via induction of strong Th2 responses, while downregulation of type 1 inflammation modifying the intestinal environment to promote their survival (Maizels et al., 2004). High expression levels of OV defence proteins, including T265_13308, in the bile duct may be responsible for immunomodulation to enable continued parasite feeding (Sithithaworn et al., 2014). The resulting immunomodulation is thought to promote helminth survival within human hosts (Suttiprapa et al., 2008).

The infiltration of inflammatory cells to sites of infection and fibrotic tissue/tumours is a consistent observation in hamster models (Sripa and Kaewkes, 2000, 2002). The infiltration of immune cells to sites of OV infection suggests that OVES products may either directly or indirectly (via influencing host cells) generate chemoattractant molecules. Neutrophils are recruited to the infected bile ducts in the early stages of infection, followed by eosinophils and monocytes which then predominate (Bhamarapravati et al., 1978; Jittimanee et al., 2007; Wongratanacheewin et al., 2003). Damage-associated molecular pattern signalling from necrotic cells damaged by OV feeding may also stimulate the recruitment of inflammatory immune cells (Kono and Rock, 2008). Much research has focussed on identifying the extracellular vesicles and soluble secreted molecules from OV in culture (OVES products) (Chaiyadet et al., 2015b, 2017). An OVES product has been hypothesized to sequester signals from necrotic cells, allowing the parasite to feed and possibly evade detection (Smout et al., 2009). This may be an example of immunomodulatory properties of OV (Wongratanacheewin et al., 1987), masking wound damage in an attempt to decrease acute immune responses. It is established that T cells are not required for the recruitment and infiltration of innate immune cells (Flavell and Flavell, 1986) but may mediate inflammatory responses within the biliary network during opisthorchiasis.

The infiltration and recruitment of inflammatory cells to the site of infection is likely to be influenced by soluble OVES molecules eliciting IL-6 and IL-8 release from cholangiocytes. These cytokines are released following Toll-like receptor (TLR)-4-mediated activation of nuclear factor kappa-light-chain enhancer of activated B cells (NF-κB) via myeloid differentiation primary response gene 88 (MyD88) (Ninlawan et al., 2010). It is likely that the production of chemoattractant (IL-8) chemokines and proinflammatory (IL-6) cytokines from these nonimmune cell cholangiocytes is driven by clathyrin-mediated endocytosis of vesicular OVES products (Chaiyadet et al., 2015a). Notably, these chemokines and cytokines released during OV infection (Jittimanee et al., 2007) are also likely to stimulate the influx of immune cells to the infection site, which then themselves generate chemoattractants. It would also be of interest to define how the complement cascade responds to OV, given that the cascade is crucial to alerting innate cells to infection and can generate its own immune cell chemoattractants, such as C5a or C3a.

2.2 Immune Cell Responses to OV/OVES Products

It is somewhat surprising that literature covering specific responses of innate inflammatory cells towards OV is rather limited, given that infiltration of these cells to the site of infection is a consistent observation in animal models of OV infection. Eosinophilia is associated with OV infection (Nishiura et al., 2003) and is common in other parasitic infections (Shin et al., 2009). In the case of OV, there is early infiltration of neutrophils and both eosinophils and macrophages are persistently abundant around the site of infections after several weeks and months (Bhamarapravati et al., 1978; Sripa and Kaewkes, 2000). Therefore, it would be important to determine the specific responses of eosinophils and macrophages to OV/OVES products. This may reveal if innate responses are regulated directly by OV/OVES products and whether such a mechanism could explain the inflammatory damage observed in opisthorchiasis.

We have made the recent and novel discovery that circulating blood neutrophils from OV-infected individuals have a greater capacity to produce reactive oxygen metabolites ROS (reactive oxygen species) and undergo phagocytosis than healthy controls (unpublished observations). These circulating neutrophils also have elevated surface expression of the integrin, CD11b. Moreover, neutrophils from OV-infected individuals with advanced periductal fibrosis have an even greater capacity to generate ROS and phagocytose and are sensitized to the stimulatory effects of OVES. This previously unrecognized involvement of circulating neutrophils during OV infection, particularly in those with liver disease, indicates a systemic activation of the innate immune system and a role for these cells in the pathological events associated with persistent inflammation.

Recently, significant increases in a number of proinflammatory and antiinflammatory cytokines were measured in isolated peripheral blood mononuclear cells (PBMCs) of non-OV-infected and OV-infected, CCA-positive patients stimulated with OVES products in vitro (Hongsrichan et al., 2014; Surapaitoon et al., 2017). Although it was not clear which cells were responsible for the cytokine release, i.e., T or B lymphocytes, monocytes or neutrophils, this research demonstrated that cytokine release from both unstimulated and stimulated PBMCs of OV-induced CCA patients was significantly elevated above levels in uninfected individuals. Whilst observations that individuals may acquire successive, repeated infections suggest inefficient innate and humoral responses that fail to clear OV (Flavell et al., 1980), this PBMC response to OV suggests a persistently activated state, increased sensitivity and perhaps even memory for responses to OV/OVES products. This may provide evidence of PBMCs from CCA patients becoming more responsive or 'primed' to OVES products, developing a form of memory that allows for a more rapid and greater release of cytokines. This 'prepared' state of PBMCs to release cytokines may regulate increased inflammatory responses to repeated infection and/or further release of OVES products, and it would be of great interest to determine if this 'primed' state also occurs in other immune cells. Alternatively, altered responses of PBMCs in CCA patients may be in response to the cancer. In the case of persistent infection, the upregulation of these cytokines (and perhaps other yet unidentified cytokines) will likely recruit immune cells and promote inflammatory responses in response to OV infection. It is therefore important to fully characterize these immune responses, including cytokine profiling, in other immune cells that are activated in OV infection, such as neutrophils and monocytes.

In the RAW264 macrophage-like cell line, OV antigen activates NF-κB-induced upregulation of inducible nitric oxide synthase (iNOS) and cyclooxygenase-2 (COX-2) through TLR-2 (Pinlaor et al., 2005). This was comparable to responses of human peripheral blood leukocytes (PBLs), which also upregulated COX-2 and manganese superoxide dismutase 2 (MnSOD-2) and downregulated catalase (CAT) (Yongvanit et al., 2012a,b). However, it should be noted that this report did not specify which immune cells in the PBL population responded to OV antigen but does suggest that immune cells that possess TLR-2 receptors may be activated by OV antigen. Understanding which TLRs can be activated by OV will enhance our understanding of OV and provide insights into the biochemical composition of the parasite/antigen/OVES products and how comparable it is to that of other helminths/pathogens. Interestingly, while TLR-4 expression was not upregulated in the RAW264 cell line, it was significantly upregulated in cholangiocytes which could be responsible for enhanced IL-6 and IL-8 expression (Ninlawan et al., 2010). In addition, the upregulation of COX-2 suggests increased production of prostaglandins that may assist immune cell infiltration by stimulating processes such as vasodilation.

Both infiltrating and tissue-resident macrophages respond to OV (Pinlaor et al., 2009). However, the upregulation of iNOS in the RAW264 macrophage cell line is somewhat surprising, given that the anti-inflammatory and proliferative M2 phenotype of macrophages has been observed after chronic OV infection (Bility and Sripa, 2014; Thanee et al., 2015); this may represent differences in functions between RAW264 - cells and human macrophages. This M2 phenotype is associated with assisting wound healing processes which may be linked to fibrosis in chronic infections and promoted by the predominant Th2 response in OV-infected hamsters (Jittimanee et al., 2007). The M2 phenotype is not normally associated with proinflammatory characteristics, nor the inflammatory damage observed in opisthorchiasis. These latter phenotypes are more typical characteristics of the proinflammatory M1 phenotype. It would be of interest to determine if OV/OVES products generated at different stages of infection promote different subsets of each macrophage phenotype. Certainly, more research is required on this topic, as the concept of macrophage polarization in disease is still not well understood and requires further understanding.

2.3 OV-Induced Oxidative/Nitrative Damage and Survival of Parasites

The production of ROS/reactive nitrogen species (RNS) during a 'respiratory burst' is a key characteristic of inflammatory responses of cells such as neutrophils and macrophages during infection (Wright et al., 2010). The presence and accumulation of 8-nitroguanine (Pinlaor et al., 2003) and 8-oxo-7,8-dihydro-2′-deoxyguanosine (8-oxodG) lesions observed in the bile ducts of hamsters with repeated OV infection may be associated with immune cell iNOS production of nitric oxide (NO) (MacMicking et al., 1997; Pinlaor et al., 2004). These DNA lesions may also be a useful predictor of CCA (Saichua et al., 2015) because DNA damage increases the risk of mutations during cell division that precedes tumour cell formation. NO may also react with hydrogen peroxide (H_2O_2) produced by SOD-2, which is upregulated in PBLs exposed to OV antigen (Yongvanit et al., 2012a,b), to form peroxynitrite ($ONOO^-$), another free radical, thereby further increasing the risk of DNA damage.

Free radicals might not just be directed towards OV. The liver fluke is also believed to be a reservoir host for the known carcinogen, *H. pylori* (Deenonpoe et al., 2015), which, if it accesses the biliary network, could exacerbate inflammatory responses (Dangtakot et al., 2017) and even promote carcinogenesis, as it does in gastric cancer (Ajani et al., 2017). Free radicals may also target invading bacteria in the altered microbiome observed during opisthorchiasis (Plieskatt et al., 2013; Itthitaetrakool et al., 2016). This may represent another layer of complexity to OV infection, whereby immune cell infiltration may be in response to bacteria that were once commensal but have become pathogenic. The alteration of the microbiome and increases in fibrotic tissue in chronic infection are believed to limit the contractility of bile ducts. This, in turn, may allow the accumulation of bile sludge (Mairiang et al., 2012) that may pose an additional threat to the DNA of local cells.

An increased expression of the antioxidant peroxiredoxin-6 in inflammatory cells during opisthorchiasis (Khoontawad et al., 2010) indicates the host's attempt to protect itself from free radicals generated during infection. However, upregulation of oxidative enzymes and downregulation of catalase in PBLs may be responsible for the overall accumulation of free radicals and DNA damage in OV infection. Whilst cells possess DNA repair mechanisms that can limit or reverse DNA damage, an increased oxidative environment may overwhelm repair mechanisms, especially if OV can alter or even initiate mutations in these repair mechanisms (Tangkawattana et al., 2008; Jusakul et al., 2015). It is evident that OV can survive inflammatory responses, although observations following treatment with antiinflammatory drugs suggest the parasite is healthier when inflammatory responses are absent (Juasook et al., 2012). OV's ability to survive in this oxidative environment may be due to the ability of thioredoxin (Trx), identified in OV and OVES products (Suttiprapa et al., 2008, 2012) to metabolize free radicals. This antioxidant may protect OV from oxidative damage. This is an example of how the parasite has adapted to protect itself from oxidative/nitrative damage, whilst local cells become damaged and hence are at risk of initiation of tumourigenesis.

3. HOW DOES OV INFECTION RESULT IN DEVELOPMENT OF CCA?

3.1 CCA: An OV-Induced Wound That Never Heals

Oxidative/nitrative damage is one mechanism by which inflammatory responses may directly initiate cancer. However, innate inflammatory responses may indirectly support tumourigenesis by promoting proliferative responses or inhibiting repair mechanisms during infection. Persistent OV infection prevents damaged cholangiocytes from fully repairing, which is a proposed characteristic of tumours (Dvorak, 1986). Usually, inflammatory cells leave sites of tissue damage or die by apoptosis once infection is cleared and then wound healing begins. However, persistent OV infection may cause inflammatory cells to remain throughout the infection (either by delayed apoptosis or by continuous rounds of infiltration), which will continue to release inflammatory and proliferative signals. The antiapoptotic activity of Trx in the OVES products has been described in cholangiocytes exposed to H_2O_2 in an oxidative stress model (Matchimakul et al., 2015). In addition to releasing free radicals and phagocytosing debris and necrotic cells, macrophages recruit fibroblasts via platelet-derived growth factor (PDGF) and transforming growth factor-beta (TGF-β); isoforms of both these cytokines are increased in response to OV infection (Smout et al., 2009; Boonjaraspinyo et al., 2012). Moreover, it is also recognized that activated neutrophils can generate locally high levels of proinflammatory and antiinflammatory cytokines and chemokines (Fig. 3), together with a range of growth factors and angiogenic mediators (Mantovani et al., 2011). High levels of IL-1β signalling (Pinlaor et al., 2005) from immune cells and increased collagen production and matrix metalloproteinases (MMPs) from recruited immune cells and fibroblasts are likely to result in an excess of extracellular matrix (ECM) required for tissue remodelling (Reviewed in Page-McCaw et al., 2007; Prakobwong et al., 2009). A role for MMPs in fibrosis-associated CCA in the hamster model has been proposed (Prakobwong et al., 2010) and MMP/TIMP

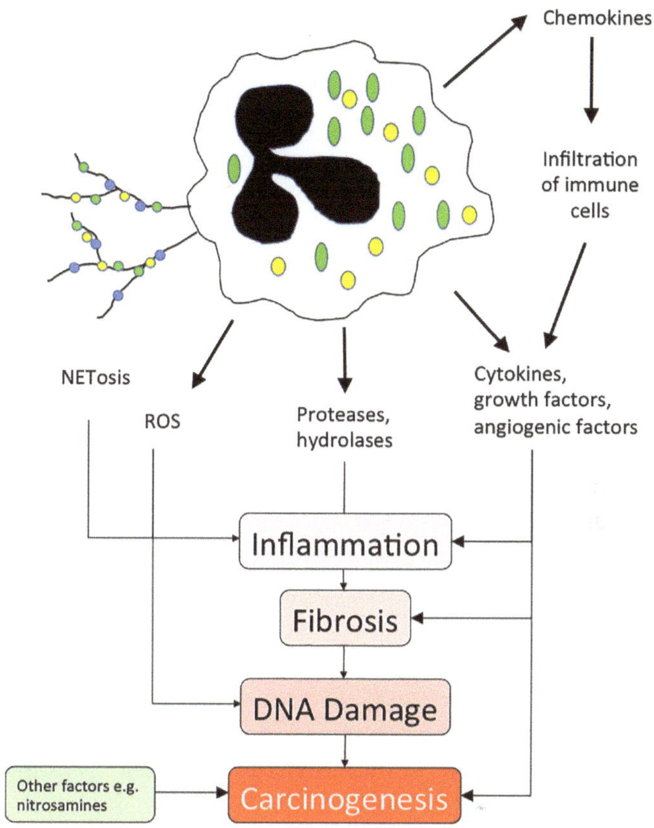

Figure 3 Neutrophil activation and generation/release of factors that can promote autoimmunity, inflammation and tissue damage and DNA damage. *ROS*, reactive oxygen species.

levels are elevated in OV-infected patients with CCA (Prakobwong et al., 2012). This continuous and excessive wound healing is likely to result in fibrotic tissue developing and promote an overproliferative phenotype in tissue cells, encouraging cells to replicate repetitively and uncontrollably. In addition, the proliferative granulin-like protein observed in OV (Smout et al., 2009) may also assist the wound healing response and tumour development.

Proliferative and proinflammatory cytokines upregulated in opisthorchiasis, such as IL-4, IL-6 and TGF-β, are required for wound healing processes, such as angiogenesis and cell proliferation. Specifically, IL-6, which is implicated in fibrosis and tumourigenesis (Fielding et al., 2014; Naugler and Karin, 2008), is associated with the development of APF and CCA in opisthorchiasis (Sripa et al., 2009, 2012b; Frampton et al., 2012). However, this association is only observed in OV-infected individuals with APF and/or CCA, not in OV-infected individuals without APF and CCA. This suggests OV may not directly drive this IL-6 production, but APF/CCA may create the environment responsible for overproduction of IL-6. This implicates IL-6 as an indicator of CCA in OV-infected patients, which may be used to predict the onset of tumour formation, and this cytokine has previously been associated with cancers (Chiu et al., 1996; Wehbe et al., 2006). IL-6 may even increase the number of macrophages at sites of infection, encouraging further inflammatory and proliferative signalling. Inflammatory responses are likely to be responsible for the accumulation of fibrotic tissue, which can be identified via ultrasound. Because most OV-infected CCA patients present with fibrosis before the onset of the cancer, this further implicates inflammatory and fibrotic responses in the development of CCA (Sripa et al., 2012a).

Once a tumour has developed, it requires its own tumour microenvironment (TME) to survive. The TME often constitutes immune cells, fibroblasts, blood vessels, inflammatory cells, growth factors, and ECM: however, the TME is only partially defined in OV-induced CCA (Utispan et al., 2010; Chuaysri et al., 2009). The continued supply of growth factors, cytokines, nutrients and oxygen and enhanced vascularization during inflammatory/proliferative responses towards persistent OV infection are likely to help tumour progression. One component of the TME is tumour-associated macrophages (TAMs), which develop the M2 phenotype via processes that may be promoted by OV. This phenotype is considered protumourigenic and may support the TME for development of CCA and may even be implicated in tumour metastasis (Bility and Sripa, 2014). As the tumour establishes, it is likely to adopt a 'mutator' phenotype, characterized by increased genomic instability and erroneous DNA repair. Inflammatory cell release of free radicals during persistent OV infection may then allow mutations in DNA to occur more frequently. The chronic inflammation-associated upregulation of proliferative signalling pathways, phosphoinositide 3-kinase (PI3K)/protein kinase B (AKT) and Wnt/β-catenin signalling in OV-induced CCA tissues (Yothaisong et al., 2014) may support the mutator phenotype, allowing mutated cells to replicate quicker and more frequently, contributing to the overproliferative and uncontrolled replication often exhibited in tumours before they metastasize.

3.2 Genetic and Epigenetic Changes and Susceptibility to CCA

Whilst it is well established that host inflammation contributes to fibrosis and CCA as a result of OV infection, not every infected individual develops CCA (Sripa et al., 2012a). Of those infected in Thailand, only ~25% develop advanced periductal fibrosis, and ~1%–2% develop CCA (Mairiang et al., 2012). Whilst these are low percentages, because of the high number of people infected, this accounts for >26,000 CCA-associated deaths per year (Sripa and Pairojkul, 2008). The molecular basis for this heterogenous disease progression following infection is unknown.

Epigenetic alterations in host genomes are capable of causing susceptibility to various cancers and immune disorders and may play a role in the development of CCA. Whole-exome sequencing of the CCA genome (Jusakul et al., 2015) revealed novel gene mutations in CCA that are involved in chromatin function, including BAP1 (2.8%), ARID1A (17.6%), MLL3 (13%) and IDH1/2 (2.8%). Decreased expression and functionality of mismatch repair gene hMLH1 were also detected in 44.6% of OV-associated CCA (Jusakul et al., 2015). The detection of DNA methylation at novel CCA-mutated genes has been suggested as a biomarker for CCA (Jusakul et al., 2015). Aberrant DNA methylation is responsible for changes in cholangiocyte cell fate, favouring malignancy (Chiang et al., 2015). The identification of novel mutations in the CCA genome could allow for targeted DNA methylation analysis, using next generation sequencing, enabling rapid diagnosis of CCA (Masser et al., 2015). In addition, if identified gene mutations follow the two-hit or three-hit hypothesis towards CCA development, individuals at high risk of CCA development could be identified via mutated gene alleles (Knudson, 1971; Segditsas et al., 2009). The use of genome-wide association studies, as used to investigate disorders in immune function, could be applied to investigate dysregulated immunity in CCA (Knight, 2013).

However, it is also important to note that where carcinogenesis arises as a result of viruses, bacteria and other helminths, carcinogenesis usually occurs following a lifetime of infection allowing genetic alterations to accumulate over a time, ultimately leading to a malignant phenotype, as opposed to cancer development in those with a preexisting genetic predisposition (Smout et al., 2011).

4. *HELICOBACTER*: A CANCER-INDUCING BACTERIUM

4.1 *Helicobacter* and Gastric Disease

Helicobacter spp., especially *H. pylori*, has long been recognized as a risk factor for gastroduodenal diseases, including gastritis, peptic ulcer, gastric cancer and mucosa-associated lymphoid tissue lymphoma (Moss, 2017; Kao et al., 2016). Around 50% of the world's population are infected, and it has been classified by the WHO as a type I carcinogen. Most infected individuals have asymptomatic gastritis, but 10%—20% have an increased risk of developing peptic ulcers, and 1%—2% infected individuals have an increased risk of developing distal gastric cancer (Moss, 2017). Infection with *H. pylori* occurs in early life, and it may be beneficial in suppressing asthma, allergies and inflammatory bowel disease (Reibman et al., 2008; Chen and Blaser, 2008; Luther et al., 2010; Koch and Muller, 2015). Host gene polymorphisms may, at least in part, account for these varied pathological outcomes following infection. For example, IL-1β polymorphisms resulting in enhanced levels of expression of this molecule may predispose to gastric cancer as this molecule inhibits acid secretion that may lead to gastric atrophy, metaplasia and then gastric cancer, as well as inducing proliferation of gastric carcinoma cells (Hwang et al., 2002). IL-1β is therefore one of the key host factors for *H. pylori*—associated gastric cancer progression.

4.2 *Helicobacter* and Inflammation

H. pylori infection is associated with a large infiltration of neutrophils (together with monocytes and some other immune cells), which leads to gastritis, and there is a strong correlation between bacterial load and neutrophil numbers/inflammation scores (Xu et al., 2012). Neutrophil inflammation resolves following successful antibiotic treatment. In addition to neutrophil infiltration, there is evidence of neutrophil-mediated ROS production and myeloperoxidase release within the gastric epithelium (Davies et al., 1994; Fu et al., 2016), and these molecules may contribute to the tissue damage and events associated with the various gastric pathologies, including gastric cancer.

H. pylori colonization and pathogenicity is dependent on a number of key genes and proteins. These are (Kao et al., 2016) as follows: (1) urease, which is responsible for the conversion of urea to ammonia, may help in generating a localized, nonacidic microenvironment around the bacteria to promote their survival in the stomach. This enzyme may also affect the pH of phagosomes following phagocytosis by neutrophils and hence interfere with bacterial killing processes; (2) a 40-kbp pathogenicity island in the more virulent strains (CagPAI) is a type IV secretory system that effectively delivers CagA protein into host cells to regulate a variety of intracellular signaling systems in the target cells; (3) vacuolating toxin A (VacA) induces toxicity in epithelial and immune cells; and (4) the bacteria also express a number of surface adhesion proteins that promote binding to gastric epithelial cells and they are highly motile.

One of the first *H. pylori* molecules to be characterized was HP-NAP (neutrophil-activating protein). It was first shown that a water-soluble molecule could be isolated from *H. pylori* that was capable of activating ROS production and chemotaxis in neutrophils (Mooney et al., 1991). This was later purified and characterized as HP-NAP, which is a 150-kDa molecule comprising 10 identical 15-kDa subunits (Evans et al., 1995). The NAP gene is present in all *Helicobacter* strains, but there is considerable heterogeneity in expression levels between strains. The 3D structure of HP-NAP resembles that of bacterioferritins (Tsuruta et al., 2012), and it is highly conserved. HP-NAP is normally present in the bacterial cytosol, and it is likely released after lysis. It is chemotactic for neutrophils, monocytes and other immune cells, and in addition to its ability to stimulate ROS and myeloperoxidase release from neutrophils (Wang et al., 2008), it can induce the secretion of a variety of cytokines and chemokines that can play a role in immune activation and pathology. For example, it stimulates the secretion of TNFα, IL-8, CCL (Chemokine C—C motif ligand) 3, CCL4 from neutrophils and IL-6 and other cytokines from monocytes (Alvarez-Arellano et al., 2007; Amedei et al., 2006). It can also trigger IL-12/23 release from neutrophils and monocytes thereby promoting a Th1 response and inhibiting a Th2 response (D'Elios and Andersen, 2007; de Bernard and D'Elios, 2010).

Other *H. pylori* molecules that can regulate cytokine/chemokine release include Hsp (Heat Shock Protein) 60 (GroEL-like HspB) and Hsp10 (GroES-like HspA) that induce IL-8, TNFα and Gro expression via TLR2-mediated activation of NF-κB (Lin et al., 2009; Zhao et al., 2007). In addition to stimulating the expression of these proinflammatory molecules, the antiinflammatory cytokine, IL-10 may also be activated (Pachathundikandi and Backert, 2018), indicating complex regulatory mechanisms to fine tune the inflammatory process following neutrophil infiltration and activation.

4.3 The Inflammasome

The inflammasome is a multicomponent protein complex that becomes assembled in innate myeloid cells in response to external signals such as pathogens and it regulates the processing and expression of inflammatory cytokines such as IL-1β and IL-18. Inflammasome formation (reviewed in Broz and Dixit, 2016) is often triggered by ligation of PRRs such as TLRs, C-type lectin receptors (CLRs) or intracellular NOD-like receptors (NLRs). Their precise structure can vary, but generally they comprise an NOD-like receptor (e.g., NLRP1, NLRP3 or NLRP4), pro-caspase-1 and apoptosis-associated speck-like protein containing a CARD (cascade activation and recruitment domain). Once assembled, pro-caspase-1 becomes activated and can cleave pro-IL-1β and pro-IL-18 to active forms, which, when released, can regulate a number of cell and tissue functions. Inflammasomes containing NLRP3 have been described following exposure of neutrophils or other innate immune cells to *H. pylori* and its products (Semper et al., 2014; Perez-Figueroa et al., 2016; Pachathundikandi and Backert, 2018).

Inflammasome formation requires ROS production, potassium efflux and lysosomal destabilization (Semper et al., 2014), and both priming and second signals are required. The processes by which *H. pylori* activates the inflammasome in human innate cells are becoming clarified, but there is some debate in the literature as to the precise mechanisms involved. In part, these differences may be explained by the different experimental systems used (e.g., blocking TLR antibodies). A consensus view is emerging that bacterial components (e.g., LPS, glycopeptides, flagellin, UreB) activate TLR2 and TLR4, which then result in NF-κB activation. This then activates transcription of NLPR3 and pro-IL-1β and pro-IL-18, and the expressed NLPR3 assembles with ASC (apoptosis-associated speck-like protein containing a CARD) and pro-caspase-1 to form the inflammasome. Activated caspase-1 can then cleave the pro-cytokines that are then released to activate immune cells and tissue cells. Experiments using *H. pylori* strains and mutants with altered expression of key components indicate that the CagPAI and VacA are required (but perhaps not CagA) for activation of inflammasome assembly in immune cells (Pachathundikandi and Backert, 2018; Semper et al., 2014).

In addition to triggering cytokine expression, ROS production and degranulation, it was recently shown that human neutrophils adopt a new phenotype, resembling the N1 phenotype (Whitmore et al., 2017) following phagocytosis of *H. pylori*. Following phagocytosis neutrophils become $CD62L^{dim}$, $CD11b^{bright}$, $CD66b^{bright}$ and $CD63^{bright}$, in addition generating proinflammatory cytokines. They also have delayed apoptosis and develop a hypersegmented nucleus (Whitmore et al., 2017). Experiments with different strains of *H. pylori* indicate that HP-NAP, VacA, CagG and CagT are not required for the neutrophils to acquire this unusual phenotype that is sometimes observed in clinical biopsies.

4.4 *Helicobacter* and Liver Fluke Infections

Helicobacter spp. (including *H. pylori*, *H. hepaticus* and *H. bilis*) are detected in the bile (Fukuda et al., 2002), gallbladder (Chen et al., 2003) and liver tissue (Huang et al., 2004) of patients with a number of hepatobiliary diseases. *H. pylori* is also detected in patients with liver carcinoma and CCA (Huang et al., 2004; Pellicano et al., 2004; Xuan et al., 2006; Abu Al-Soud et al., 2008).

A systematic study of 87 patients with liver fluke—associated CCA, 53 with cholelithiasis (benign disease group) and healthy control specimens (from autopsies) identified *H. pylori* in 66.7%, 41.5% and 25.0% of patients in CCA, cholelithiasis and control groups ($P < .05$), respectively (Boonyanugomol et al., 2012). *H. bilis* was detected in 14.9% and 9.4% of the patients with CCA and cholelithiasis, respectively ($P < .05$), but this species was absent from the control group. The *cagA* gene was detected in 36.2% and 9.1% of patients with CCA and cholelithiasis, respectively ($P < .05$, Boonyanugomol et al., 2012).

Further links between *H. pylori* and OV infection have come from animal studies. For example, coinfection of Syrian hamsters with OV and *H. pylori* resulted in the most severe hepatobiliary abnormalities, including periductal fibrosis, cholangitis and bile duct hyperplasia and hence a lower survival rate, compared with animals infected with either pathogen alone or control groups (Dangtakot et al., 2017). The highest levels of IL-1, IL-6 and TNFα in liver tissues were also observed in the dual-infected group. A microbiome/metagenomic study of hamsters infected with OV identified several bacterial species in the livers of infected hamsters, including *H. pylori* and other *Helicobacter* spp. (Itthitaetrakool et al., 2016). There are suggestions that OV may serve as a reservoir for *H. pylori* (Deenonpoe et al., 2015), which could explain the presence of this cancer-inducing bacterium following OV infection. A recent large-scale study of 500 individuals showed that liver fluke infection was associated with a higher frequency of CagA-positive *H. pylori*, present in 64.6% of stool samples compared with a presence of only 29.6% in stool samples from uninfected controls (Deenonpoe et al., 2017). The CagPAI is also necessary for bacterial internalization in primary hepatobiliary cells derived from CCA patients (Boonyanugomol et al., 2013).

4.5 OV and Activation of Innate Immunity: The *Helicobacter* Link

The recent discovery of a link between *H. pylori* coinfection with OV infection has a number of implications for understanding disease pathology and possible new avenues for treatment (Sripa et al., 2017). It is tempting to speculate that *Helicobacter* coinfection can stimulate the innate immune system, by similar mechanisms that occur in gastric disease, to activate the immune system in opisthorchiasis and contribute to the events that lead to periductal fibrosis and CCA. *H. pylori*—derived factors may be responsible for directly or indirectly activating neutrophils in OV-infected individuals, which may contribute to the release of tissue-damaging molecules and generation of proinflammatory cytokines and chemokines.

5. CONCLUDING REMARKS AND FUTURE PERSPECTIVES

OV has clear carcinogenic potential. Whether this is a direct result of promoting overinflammatory conditions, overproliferative responses or decreased repair mechanisms, CCA onset is triggered by OV. The literature to date indicates that various innate immune responses that are predicted to occur or have been measured during OV-induced inflammation can lead to cell damage that may initiate or trigger carcinogenesis. Whether OV directly causes cancer, or whether the hypothesized 'proinflammatory phenotype' is directly responsible for the onset of cancer or produces an environment for cancer to develop from other carcinogens, is still to be determined. Also, it is necessary to determine the role played by *Helicobacter* spp. in the events that lead to the development of CCA. In particular, it is necessary to determine if this bacterium directly activates the innate immune system in similar ways that it does in gastric diseases. It would therefore be of interest to determine if anti-*Helicobacter* antibiotic treatment will be beneficial in preventing the development of CCA following OV infection. It is possible that OV has evolved mechanisms to generate the inflammatory environment and cancer so that it then has access to a greater source of nutrients. However, the parasite itself is largely resistant to the attempts of the innate immune system to destroy and eliminate it. More research is required before the proinflammatory phenotype can be fully characterized. However, if this phenotype can be demonstrated, it may allow a better understanding of how infectious agents can drive inflammation-related cancer and perhaps other inflammatory diseases. This may lead to the development of new therapeutic approaches perhaps drawing on analogies in diseases such as rheumatoid arthritis, where targeting of the immune system with specific biologic drugs lead to dramatic improvement in disease progression.

REFERENCES

Abu Al-Soud, W., Stenram, U., Ljungh, A., Tranberg, K.G., Nilsson, H.O., Wadstrom, T., 2008. DNA of *Helicobacter* spp. and common gut bacteria in primary liver carcinoma. Dig. Liver Dis. 40, 126–131.

Ajani, J.A., Lee, J., Sano, T., Janjigian, Y.Y., Fan, D.M., Song, S.M., 2017. Gastric adenocarcinoma. Nat. Rev. Dis. Primers 3. https://doi.org/10.1038/nrdp.2017.36.

Alvarez-Arellano, L., Camorlinga-Ponce, M., Maldonado-Bernal, C., Torres, J., 2007. Activation of human neutrophils with *Helicobacter pylori* and the role of Toll-like receptors 2 and 4 in the response. FEMS Immunol. Med. Microbiol. 51, 473–479.

Amedei, A., Cappon, A., Codolo, G., Cabrelle, A., Polenghi, A., Benagiano, M., Tasca, E., Azzurri, A., D'Elios, M.M., Del Prete, G., de Bernard, M., 2006. The neutrophil-activating protein of *Helicobacter pylori* promotes Th1 immune responses. J. Clin. Invest. 116, 1092–1101.

Andrews, R.H., Sithithaworn, P., Petney, T.N., 2008. *Opisthorchis viverrini*: an underestimated parasite in world health. Trends Parasitol. 24, 497–501.

Balkwill, F., Mantovani, A., 2001. Inflammation and cancer: back to Virchow? Lancet 357, 539–545.

Bartsch, H., Montesano, R., 1984. Relevance of nitrosamines to human cancer. Carcinogenesis 5, 1381–1393.

Bhamarapravati, N., Thammavit, W., Vajrasthira, S., 1978. Liver changes in hamsters infected with a liver fluke of man, *Opisthorchis viverrini*. Am. J. Trop. Med. Hyg. 27, 787–794.

Bility, M.T., Sripa, B., 2014. Chronic *Opisthorchis viverrini* infection and associated hepatobiliary disease is associated with iron loaded M2-like macrophages. Kor. J. Parasitol. 52, 695–699.

Boonjaraspinyo, S., Wu, Z., Boonmars, T., Kaewkes, S., Loilome, W., Sithithaworn, P., Nagano, I., Takahashi, Y., Yongvanit, P., Bhudhisawasdi, V., 2012. Overexpression of PDGFA and its receptor during carcinogenesis of *Opisthorchis viverrini*-associated cholangiocarcinoma. Parasitol. Int. 61, 145–150.

Boonmars, T., Wu, Z., Boonjaruspinyo, S., Puapairoj, A., Kaewsamut, B., Nagano, I., Pinlaor, S., Yongvanit, P., Wonkchalee, O., Juasook, A., Sudsarn, P., Srisawangwong, T., 2011. Involvement of c-Ski oncoprotein in carcinogenesis of cholangiocarcinoma induced by *Opisthorchis viverrini* and N-nitrosodimethylamine. Pathol. Oncol. Res. 17, 219–227.

Boonyanugomol, W., Chomvarin, C., Hahnvajanawong, C., Sripa, B., Kaparakis-Liaskos, M., Ferrero, R.L., 2013. *Helicobacter pylori* cag pathogenicity island (cagPAI) involved in bacterial internalization and IL-8 induced responses via NOD1- and MyD88-dependent mechanisms in human biliary epithelial cells. PLoS One 8, e77358.

Boonyanugomol, W., Chomvarin, C., Sripa, B., Bhudhisawasdi, V., Khuntikeo, N., Hahnvajanawong, C., Chamsuwan, A., 2012. *Helicobacter pylori* in Thai patients with cholangiocarcinoma and its association with biliary inflammation and proliferation. HPB 14, 177–184.

Broz, P., Dixit, V.M., 2016. Inflammasomes: mechanism of assembly, regulation and signaling. Nat. Rev. Immunol. 16, 407–420.

Chaiyadet, S., Smout, M., Johnson, M., Whitchurch, C., Turnbull, L., Kaewkes, S., Sotillo, J., Loukas, A., Sripa, B., 2015a. Excretory/secretory products of the carcinogenic liver fluke are endocytosed by human cholangiocytes and drive cell proliferation and IL6 production. Int. J. Parasitol. 45, 773–781.
Chaiyadet, S., Smout, M., Laha, T., Sripa, B., Loukas, A., Sotillo, J., 2017. Proteomic characterization of the internalization of *Opisthorchis viverrini* excretory/secretory products in human cells. Parasitol. Int. 66, 494–502.
Chaiyadet, S., Sotillo, J., Smout, M., Cantacessi, C., Jones, M.K., Johnson, M.S., Turnbull, L., Whitchurch, C.B., Potriquet, J., Laohaviroj, M., Mulvenna, J., Brindley, P.J., Bethony, J.M., Laha, T., Sripa, B., Loukas, A., 2015b. Carcinogenic liver fluke secretes extracellular vesicles that promote cholangiocytes to adopt a tumorigenic phenotype. J. Infect. Dis. 212, 1636–1645.
Chen, W., Li, D., Cannan, R.J., Stubbs, R.S., 2003. Common presence of *Helicobacter* DNA in the gallbladder of patients with gallstone diseases and controls. Dig. Liver Dis. 35, 237–243.
Chen, Y.J., Chang, Y.T., Wang, C.B., Wu, C.Y., 2011. The risk of cancer in patients with rheumatoid arthritis: a nationwide cohort study in Taiwan. Arthritis Rheum. 63, 352–358.
Chen, Y., Blaser, M.J., 2008. *Helicobacter pylori* colonization is inversely associated with childhood asthma. J. Infect. Dis. 198, 553–560.
Chiang, N.J., Shan, Y.S., Hung, W.C., Chen, L.T., 2015. Epigenetic regulation in the carcinogenesis of cholangiocarcinoma. Int. J. Biochem. Cell Biol. 67, 110–114.
Chiu, J.J., Sgagias, M.K., Cowan, K.H., 1996. Interleukin 6 acts as a paracrine growth factor in human mammary carcinoma cell lines. Clin. Canc. Res. 2, 215–221.
Chuaysri, C., Thuwajit, P., Paupairoj, A., Chau-In, S., Suthiphongchai, T., Thuwajit, C., 2009. Alpha-smooth muscle actin-positive fibroblasts promote biliary cell proliferation and correlate with poor survival in cholangiocarcinoma. Oncol. Rep. 21, 957–969.
D'Elios, M.M., Andersen, L.P., 2007. *Helicobacter pylori* inflammation, immunity, and vaccines. Helicobacter 12, 15–19.
Dangtakot, R., Pinlaor, S., Itthitaetrakool, U., Chaidee, A., Chomvarin, C., Sangka, A., Wilailuckana, C., Pinlaor, P., 2017. Coinfection with *Helicobacter pylori* and *Opisthorchis viverrini* enhances the severity of hepatobiliary abnormalities in hamsters. Infect. Immun. 85, e00009–e00017.
Davies, G.R., Simmonds, N.J., Stevens, T.R., Sheaff, M.T., Banatvala, N., Laurenson, I.F., Blake, D.R., Rampton, D.S., 1994. *Helicobacter pylori* stimulates antral mucosal reactive oxygen metabolite production in vivo. Gut 35, 179–185.
de Bernard, M., D'Elios, M.M., 2010. The immune modulating activity of the *Helicobacter pylori* HP-NAP: friend or foe? Toxicon 56, 1186–1192.
Deenonpoe, R., Chomvarin, C., Pairojkul, C., Chamgramol, Y., Loukas, A., Brindley, P.J., Sripa, B., 2015. The carcinogenic liver fluke *Opisthorchis viverrini* is a reservoir for species of Helicobacter. Asian Pac. J. Cancer Prev. APJCP 16, 1751–1758.
Deenonpoe, R., Mairiang, E., Mairiang, P., Pairojkul, C., Chamgramol, Y., Rinaldi, G., Loukas, A., Brindley, P.J., Sripa, B., 2017. Elevated prevalence of *Helicobacter* species and virulence factors in opisthorchiasis and associated hepatobiliary disease. Sci. Rep. 7, 42744.
Dvorak, H.F., 1986. Tumors: wounds that do not heal. Similarities between tumor stroma generation and wound healing. N. Engl. J. Med. 315, 1650–1659.
Evans, D.J., Evans, D.G., Takemura, T., Nakano, H., Lampert, H.C., Graham, D.Y., Granger, D.N., Kvietys, P.R., 1995. Characterization of a *Helicobacter-pylori* neutrophil-activating protein. Infect. Immun. 63, 2213–2220.

Fielding, C.A., Jones, G.W., McLoughlin, R.M., McLeod, L., Hammond, V.J., Uceda, J., Williams, A.S., Lambie, M., Foster, T.L., Liao, C.T., Rice, C.M., Greenhill, C.J., Colmont, C.S., Hams, E., Coles, B., Kift-Morgan, A., Newton, Z., Craig, K.J., Williams, J.D., Williams, G.T., Davies, S.J., Humphreys, I.R., O'Donnell, V.B., Taylor, P.R., Jenkins, B.J., Topley, N., Jones, S.A., 2014. Interleukin-6 signaling drives fibrosis in unresolved inflammation. Immunity 40, 40–50.

Flavell, D.J., Flavell, S.U., 1986. *Opisthorchis viverrini*: pathogenesis of infection in immunodeprived hamsters. Parasite Immunol. 8, 455–466.

Flavell, D.J., Pattanapanyasat, K., Flavell, S.U., 1980. *Opisthorchis viverrini*: partial success in adoptively transferring immunity with spleen cells and serum in the hamster. J. Helminthol. 54, 191–197.

Frampton, G., Invernizzi, P., Bernuzzi, F., Pae, H.Y., Quinn, M., Horvat, D., Galindo, C., Huang, L., McMillin, M., Cooper, B., Rimassa, L., DeMorrow, S., 2012. Interleukin-6-driven progranulin expression increases cholangiocarcinoma growth by an Akt-dependent mechanism. Gut 61, 268–277.

Fu, H., Ma, Y., Yang, M., Zhang, C., Huang, H., Xia, Y., Lu, L., Jin, W., Cui, D., 2016. Persisting and increasing neutrophil infiltration associates with gastric carcinogenesis and E-cadherin downregulation. Sci. Rep. 6, 29762.

Fukuda, K., Kuroki, T., Tajima, Y., Tsuneoka, N., Kitajima, T., Matsuzaki, S., Furui, J., Kanematsu, T., 2002. Comparative analysis of *Helicobacter* DNAs and biliary pathology in patients with and without hepatobiliary cancer. Carcinogenesis 23, 1927–1931.

Ghouri, Y.A., Mian, I., Blechacz, B., 2015. Cancer review: cholangiocarcinoma. J. Carcinog. 14, 1. https://doi.org/10.4103/1477-3163.151940.

Grivennikov, S.I., Greten, F.R., Karin, M., 2010. Immunity, inflammation, and cancer. Cell 140, 883–899.

Hongsrichan, N., Intuyod, K., Pinlaor, P., Khoontawad, J., Yongvanit, P., Wongkham, C., Roytrakule, S., Pinlaora, S., 2014. Cytokine/chemokine secretion and proteomic identification of upregulated annexin A1 from peripheral blood mononuclear cells cocultured with the liver fluke *Opisthorchis viverrini*. Infect. Immun. 82, 2135–2147.

Huang, Y., Fan, X.G., Wang, Z.M., Zhou, J.H., Tian, X.F., Li, N., 2004. Identification of helicobacter species in human liver samples from patients with primary hepatocellular carcinoma. J. Clin. Pathol. 57, 1273–1277.

Hwang, I.R., Kodama, T., Kikuchi, S., Sakai, K., Peterson, L.E., Graham, D.Y., Yamaoka, Y., 2002. Effect of interleukin 1 polymorphisms on gastric mucosal interleukin 1 beta production in *Helicobacter pylori* infection. Gastroenterology 123, 1793–1803.

IARC, 2011. IARC. Monographs on the Evaluation of Carcinogenic Risks to Humans, Volume 100. A Review of Carcinogen—Part B: Biological Agents. International Agency for Research on Cancer, Lyon.

Itthitaetrakool, U., Pinlaor, P., Pinlaor, S., Chomvarin, C., Dangtakot, R., Chaidee, A., Wilailuckana, C., Sangka, A., Lulitanond, A., Yongvanit, P., 2016. Chronic *Opisthorchis viverrini* infection changes the liver microbiome and promotes *Helicobacter* growth. PLoS One 11, e0165798.

Jittimanee, J., Sermswan, R.W., Puapairoj, A., Maleewong, W., Wongratanacheewin, S., 2007. Cytokine expression in hamsters experimentally infected with *Opisthorchis viverrini*. Parasite Immunol. 29, 159–167.

Juasook, A., Boonmars, T., Kaewkes, S., Loilome, W., Veteewuthacharn, K., Wu, Z., Yongvanit, P., 2012. Anti-inflammatory effect of prednisolone on the growth of human liver fluke in experimental opisthorchiasis. Parasitol. Res. 110, 2271–2279.

Jusakul, A., Kongpetch, S., Teh, B.T., 2015. Genetics of *Opisthorchis viverrini*-related cholangiocarcinoma. Curr. Opin. Gastroenterol. 31, 258–263.

Kao, C.Y., Sheu, B.S., Wu, J.J., 2016. *Helicobacter pylori* infection: an overview of bacterial virulence factors and pathogenesis. Biomed. J. 39, 14–23.

Khoontawad, J., Wongkham, C., Hiraku, Y., Yongvanit, P., Prakobwong, S., Boonmars, T., Pinlaor, P., Pinlaor, S., 2010. Proteomic identification of peroxiredoxin 6 for host defence against *Opisthorchis viverrini* infection. Parasite Immunol. 32, 314–323.

Knight, J.C., 2013. Genomic modulators of the immune response. Trends Genet. 29, 74–83.

Knudson Jr., A.G., 1971. Mutation and cancer: statistical study of retinoblastoma. Proc. Natl. Acad. Sci. U. S. A. 68, 820–823.

Koch, K.N., Muller, A., 2015. *Helicobacter pylori* activates the TLR2/NLRP3/caspase-1/IL-18 axis to induce regulatory T-cells, establish persistent infection and promote tolerance to allergens. Gut Microb. 6, 382–387.

Kono, H., Rock, K.L., 2008. How dying cells alert the immune system to danger. Nature Rev. Immunol. 8, 279–289.

Lim, J.H., 2011. Liver flukes: the malady neglected. Korean J. Radiol. 12, 269–279.

Lin, C.Y., Huang, Y.S., Li, C.H., Hsieh, Y.T., Tsai, N.M., He, P.J., Hsu, W.T., Yeh, Y.C., Chiang, F.H., Wu, M.S., Chang, C.C., Liao, K.W., 2009. Characterizing the polymeric status of *Helicobacter pylori* heat shock protein 60. Biochem. Biophys. Res. Commun. 388, 283–289.

Lopez-Novoa, J.M., Nieto, M.A., 2009. Inflammation and EMT: an alliance towards organ fibrosis and cancer progression. EMBO Mol. Med. 1, 303–314.

Luther, J., Dave, M., Higgins, P.D., Kao, J.Y., 2010. Association between *Helicobacter pylori* infection and inflammatory bowel disease: a meta-analysis and systematic review of the literature. Inflamm. Bowel Dis. 16, 1077–1084.

Luvira, V., Nilprapha, K., Bhudhisawasdi, V., Pugkhem, A., Chamadol, N., Kamsa-ard, S., 2016. Cholangiocarcinoma patient outcome in Northeastern Thailand: single-center prospective study. Asian Pac. J. Cancer Prev. APJCP 17, 401–406.

MacMicking, J., Xie, Q.W., Nathan, C., 1997. Nitric oxide and macrophage function. Annu. Rev. Immunol. 15, 323–350.

Mairiang, E., Laha, T., Bethony, J.M., Thinkhamrop, B., Kaewkes, S., Sithithaworn, P., Tesana, S., Loukas, A., Brindley, P.J., Sripa, B., 2012. Ultrasonography assessment of hepatobiliary abnormalities in 3359 subjects with *Opisthorchis viverrini* infection in endemic areas of Thailand. Parasitol. Int. 61, 208–211.

Maizels, R.M., Yazdanbakhsh, M., 2003. Immune regulation by helminth parasites: cellular and molecular mechanisms. Nat. Rev. Immunol. 3, 733–744.

Maizels, R.M., Balic, A., Gomez-Escobar, N., Nair, M., Taylor, M.D., Allen, J.E., 2004. Helminth parasites—masters of regulation. Immunol. Rev. 201, 89–116.

Maizels, R.M., McSorley, H.J., Smyth, D.J., 2014. Helminths in the hygiene hypothesis: sooner or later? Clin. Exp. Immunol. 177, 38–46.

Mantovani, A., Cassatella, M.A., Costantini, C., Jaillon, S., 2011. Neutrophils in the activation and regulation of innate and adaptive immunity. Nat. Rev. Immunol. 11, 519–531.

Masser, D.R., Stanford, D.R., Freeman, W.M., 2015. Targeted DNA methylation analysis by next-generation sequencing. JoVE 96. https://doi.org/10.3791/52488.

Matchimakul, P., Rinaldi, G., Suttiprapa, S., Mann, V.H., Popratiloff, A., Laha, T., Pimenta, R.N., Cochran, C.J., Kaewkes, S., Sripa, B., Brindley, P.J., 2015. Apoptosis of cholangiocytes modulated by thioredoxin of carcinogenic liver fluke. Int. J. Biochem. Cell Biol. 65, 72–80.

McSorley, H.J., Hewitson, J.P., Maizels, R.M., 2013. Immunomodulation by helminth parasites: defining mechanisms and mediators. Int. J. Parasitol. 43, 301–310.

Mooney, C., Keenan, J., Munster, D., Wilson, I., Allardyce, R., Bagshaw, P., Chapman, B., Chadwick, V., 1991. Neutrophil activation by *Helicobacter pylori*. Gut 32, 853–857.

Moss, S.F., 2017. The clinical evidence linking *Helicobacter pylori* to gastric cancer. Cell Mol. Gastroenterol. Hepatol. 3, 183–191.

Murata, M., Thanan, R., Ma, N., Kawanishi, S., 2012. Role of nitrative and oxidative DNA damage in inflammation-related carcinogenesis. J. Biomed. Biotechnol. 2012, 623019.

Naugler, W.E., Karin, M., 2008. The wolf in sheep's clothing: the role of interleukin-6 in immunity, inflammation and cancer. Trends Mol. Med. 14, 109–119.

Ninlawan, K., O'Hara, S.P., Splinter, P.L., Yongvanit, P., Kaewkes, S., Surapaitoon, A., LaRusso, N.F., Sripa, B., 2010. *Opisthorchis viverrini* excretory/secretory products induce toll-like receptor 4 upregulation and production of interleukin 6 and 8 in cholangiocyte. Parasitol. Int. 59, 616–621.

Nishiura, H., Tsunoda, T., Akao, N., 2003. A case of severe eosinophilia in the acute phase of *Opisthorchis viverrini* infection. Kansenshogaku Zasshi 77, 677–681.

Pachathundikandi, S.K., Backert, S., 2018. *Helicobacter pylori* controls NLRP3 expression by regulating hsa-miR-223-3p and IL-10 in cultured and primary human immune cells. Innate Immun. 24, 11–23.

Page-McCaw, A., Ewald, A.J., Werb, Z., 2007. Matrix metalloproteinases and the regulation of tissue remodeling. Nat. Rev. Mol. Cell Biol. 8, 221–233.

Pellicano, R., Mazzaferro, V., Grigioni, W.F., Cutufia, M.A., Fagoonee, S., Silengo, L., Rizzetto, M., Ponzetto, A., 2004. *Helicobacter* species sequences in liver samples from patients with and without hepatocellular carcinoma. World J. Gastroenterol. 10, 598–601.

Perez-Figueroa, E., Torres, J., Sanchez-Zauco, N., Contreras-Ramos, A., Alvarez-Arellano, L., Maldonado-Bernal, C., 2016. Activation of NLRP3 inflammasome in human neutrophils by *Helicobacter pylori* infection. Innate Immun. 22, 103–112.

Perrigoue, J.G., Marshall, F.A., Artis, D., 2008. On the hunt for helminths: innate immune cells in the recognition and response to helminth parasites. Cell Microbiol. 10, 1757–1764.

Pinlaor, P., Kaewpitoon, N., Laha, T., Sripa, B., Kaewkes, S., Morales, M.E., Mann, V.H., Parriott, S.K., Suttiprapa, S., Robinson, M.W., To, J., Dalton, J.P., Loukas, A., Brindley, P.J., 2009. Cathepsin F cysteine protease of the human liver fluke, *Opisthorchis viverrini*. PLoS Neglected Trop. Dis. 3, e398.

Pinlaor, S., Ma, N., Hiraku, Y., Yongvanit, P., Semba, R., Oikawa, S., Murata, M., Sripa, B., Sithithaworn, P., Kawanishi, S., 2004. Repeated infection with *Opisthorchis viverrini* induces accumulation of 8-nitroguanine and 8-oxo-7,8-dihydro-2'-deoxyguanine in the bile duct of hamsters via inducible nitric oxide synthase. Carcinogenesis 25, 1535–1542.

Pinlaor, S., Tada-Oikawa, S., Hiraku, Y., Pinlaor, P., Ma, N., Sithithaworn, P., Kawanishi, S., 2005. *Opisthorchis viverrini* antigen induces the expression of Toll-like receptor 2 in macrophage RAW cell line. Int. J. Parasitol. 35, 591–596.

Pinlaor, S., Yongvanit, P., Hiraku, Y., Ma, N., Semba, R., Oikawa, S., Murata, M., Sripa, B., Sithithaworn, P., Kawanishi, S., 2003. 8-nitroguanine formation in the liver of hamsters infected with *Opisthorchis viverrini*. Biochem. Biophys. Res. Commun. 309, 567–571.

Pittet, M.J., Swirski, F.K., 2011. Monocytes link atherosclerosis and cancer. Eur. J. Immunol. 41, 2519–2522.

Plieskatt, J.L., Deenonpoe, R., Mulvenna, J.P., Krause, L., Sripa, B., Bethony, J.M., Brindley, P.J., 2013. Infection with the carcinogenic liver fluke *Opisthorchis viverrini* modifies intestinal and biliary microbiome. FASEB J. 27, 4572–4584.

Prakobwong, S., Pinlaor, S., Yongvanit, P., Sithithaworn, P., Pairojkul, C., Hiraku, Y., 2009. Time profiles of the expression of metalloproteinases, tissue inhibitors of metalloproteases, cytokines and collagens in hamsters infected with *Opisthorchis viverrini* with special reference to peribiliary fibrosis and liver injury. Int. J. Parasitol. 39, 825–835.

Prakobwong, S., Yongvanit, P., Hiraku, Y., Pairojkul, C., Sithithaworn, P., Pinlaor, P., Pinlaor, S., 2010. Involvement of MMP-9 in peribiliary fibrosis and cholangiocarcinogenesis via Rac1-dependent DNA damage in a hamster model. Int. J. Canc. 127, 2576–2587.

Prakobwong, S., Charoensuk, L., Hiraku, Y., Pinlaor, P., Pairojkul, C., Mairiang, E., Sithithaworn, P., Yongvanit, P., Khuntikeo, N., Pinlaor, S., 2012. Plasma hydroxyproline, MMP-7 and collagen I as novel predictive risk markers of hepatobiliary disease-associated cholangiocarcinoma. Int. J. Canc. 131, E416–E424.

Reibman, J., Marmor, M., Filner, J., Fernandez-Beros, M.E., Rogers, L., Perez-Perez, G.I., Blaser, M.J., 2008. Asthma is inversely associated with *Helicobacter pylori* status in an urban population. PLoS One 3, e4060.

Saichua, P., Yakovleva, A., Kamamia, C., Jariwala, A.R., Sithithaworn, J., Sripa, B., Brindley, P.J., Laha, T., Mairiang, E., Pairojkul, C., Khuntikeo, N., Mulvenna, J., Sithithaworn, P., Bethony, J.M., 2015. Levels of 8-OxodG predict hepatobiliary pathology in *Opisthorchis viverrini* endemic settings in Thailand. PLoS Neglected Trop. Dis. 9, e0003949.

Segditsas, S., Rowan, A.J., Howarth, K., Jones, A., Leedham, S., Wright, N.A., Gorman, P., Chambers, W., Domingo, E., Roylance, R.R., Sawyer, E.J., Sieber, O.M., Tomlinson, I.P., 2009. APC and the three-hit hypothesis. Oncogene 28, 146–155.

Semper, R.P., Mejias-Luque, R., Gross, C., Anderl, F., Muller, A., Vieth, M., Busch, D.H., Prazeres da Costa, C., Ruland, J., Gross, O., Gerhard, M., 2014. *Helicobacter pylori*-induced IL-1beta secretion in innate immune cells is regulated by the NLRP3 inflammasome and requires the cag pathogenicity island. J. Immunol. 193, 3566–3576.

Shin, M.H., Lee, Y.A., Min, D.Y., 2009. Eosinophil-mediated tissue inflammatory responses in helminth infection. Kor. J. Parasitol. 47 (Suppl.), S125–S131.

Sithithaworn, P., Andrews, R.H., Nguyen, V.D., Wongsaroj, T., Sinuon, M., Odermatt, P., Nawa, Y., Liang, S., Brindley, P.J., Sripa, B., 2012a. The current status of opisthorchiasis and clonorchiasis in the Mekong Basin. Parasitol. Int. 61, 10–16.

Sithithaworn, P., Andrews, R.H., Petney, T.N., Saijuntha, W., Laoprom, N., 2012b. The systematics and population genetics of *Opisthorchis viverrini sensu lato*: implications in parasite epidemiology and bile duct cancer. Parasitol. Int. 61, 32–37.

Sithithaworn, P., Yongvanit, P., Duenngai, K., Kiatsopit, N., Pairojkul, C., 2014. Roles of liver fluke infection as risk factor for cholangiocarcinoma. J Hepatobiliary Pancreat. Sci. 21, 301–308.

Smout, M.J., Laha, T., Mulvenna, J., Sripa, B., Suttiprapa, S., Jones, A., Brindley, P.J., Loukas, A., 2009. A granulin-like growth factor secreted by the carcinogenic liver fluke, *Opisthorchis viverrini*, promotes proliferation of host cells. PLoS Pathog. 5, e1000611.

Smout, M.J., Sripa, B., Laha, T., Mulvenna, J., Gasser, R.B., Young, N.D., Bethony, J.M., Brindley, P.J., Loukas, A., 2011. Infection with the carcinogenic human liver fluke, *Opisthorchis viverrini*. Mol. Biosyst. 7, 1367–1375.

Sripa, B., Kaewkes, S., 2000. Localisation of parasite antigens and inflammatory responses in experimental opisthorchiasis. Int. J. Parasitol. 30, 735–740.

Sripa, B., Kaewkes, S., 2002. Gall bladder and extrahepatic bile duct changes in *Opisthorchis viverrini*-infected hamsters. Acta Trop. 83, 29–36.

Sripa, B., Pairojkul, C., 2008. Cholangiocarcinoma: lessons from Thailand. Curr. Opin. Gastroenterol. 24, 349–356.

Sripa, B., Brindley, P.J., Mulvenna, J., Laha, T., Smout, M.J., Mairiang, E., Bethony, J.M., Loukas, A., 2012a. The tumorigenic liver fluke *Opisthorchis viverrini*—multiple pathways to cancer. Trends Parasitol. 28, 395–407.

Sripa, B., Deenonpoe, R., Brindley, P.J., 2017. Co-infections with liver fluke and Helicobacter species: a paradigm change in pathogenesis of opisthorchiasis and cholangiocarcinoma? Parasitol. Int. 66, 383–389.

Sripa, B., Kaewkes, S., Sithithaworn, P., Mairiang, E., Laha, T., Smout, M., Pairojkul, C., Bhudhisawasdi, V., Tesana, S., Thinkamrop, B., Bethony, J.M., Loukas, A., Brindley, P.J., 2007. Liver fluke induces cholangiocarcinoma. PLoS Med. 4, e201.

Sripa, B., Mairiang, E., Thinkhamrop, B., Laha, T., Kaewkes, S., Sithithaworn, P., Tessana, S., Loukas, A., Brindley, P.J., Bethony, J.M., 2009. Advanced periductal fibrosis from infection with the carcinogenic human liver fluke *Opisthorchis viverrini* correlates with elevated levels of interleukin-6. Hepatology 50, 1273–1281.

Sripa, B., Tangkawattana, S., Laha, T., Kaewkes, S., Mallory, F.F., Smith, J.F., Wilcox, B.A., 2015. Toward integrated opisthorchiasis control in northeast Thailand: the Lawa project. Acta Trop. 141, 361–367.

Sripa, B., Thinkhamrop, B., Mairiang, E., Laha, T., Kaewkes, S., Sithithaworn, P., Periago, M.V., Bhudhisawasdi, V., Yonglitthipagon, P., Mulvenna, J., Brindley, P.J., Loukas, A., Bethony, J.M., 2012b. Elevated plasma IL-6 associates with increased risk of advanced fibrosis and cholangiocarcinoma in individuals infected by *Opisthorchis viverrini*. PLoS Neglected Trop. Dis. 6, e1654.

Surapaitoon, A., Suttiprapa, S., Khuntikeo, N., Pairojkul, C., Sripa, B., 2017. Cytokine profiles in *Opisthorchis viverrini* stimulated peripheral blood mononuclear cells from cholangiocarcinoma patients. Parasitol. Int. 66, 889–892.

Suttiprapa, S., Loukas, A., Laha, T., Wongkham, S., Kaewkes, S., Gaze, S., Brindley, P.J., Sripa, B., 2008. Characterization of the antioxidant enzyme, thioredoxin peroxidase, from the carcinogenic human liver fluke, *Opisthorchis viverrini*. Mol. Biochem. Parasitol. 160, 116–122.

Suttiprapa, S., Matchimakul, P., Loukas, A., Laha, T., Wongkham, S., Kaewkes, S., Brindley, P.J., Sripa, B., 2012. Molecular expression and enzymatic characterization of thioredoxin from the carcinogenic human liver fluke *Opisthorchis viverrini*. Parasitol. Int. 61, 101–106.

Tangkawattana, S., Kaewkes, S., Pairojkul, C., Tangkawattana, P., Sripa, B., 2008. Mutations of KRAS and TP53 in a minor proportion of *Opisthorchis viverrini*-associated cholangiocarcinomas in a hamster model. Asian Pac. J. Cancer Prev. APJCP 9, 101–106.

Thamavit, W., Bhamarapravati, N., Sahaphong, S., Vajrasthira, S., Angsubhakorn, S., 1978. Effects of dimethylnitrosamine on induction of cholangiocarcinoma in *Opisthorchis viverrini*-infected Syrian golden hamsters. Cancer Res. 38, 4634–4639.

Thanee, M., Loilome, W., Techasen, A., Namwat, N., Boonmars, T., Pairojkul, C., Yongvanit, P., 2015. Quantitative changes in tumor-associated M2 macrophages characterize cholangiocarcinoma and their association with metastasis. Asian Pac. J. Cancer Prev. APJCP 16, 3043–3050.

Tsuruta, O., Yokoyama, H., Fujii, S., 2012. A new crystal lattice structure of *Helicobacter pylori* neutrophil-activating protein (HP-NAP). Acta Crystallogr. Sect. F Struct. Biol. Cryst. Commun. 68, 134–140.

Utispan, K., Thuwajit, P., Abiko, Y., Charngkaew, K., Paupairoj, A., Chau-in, S., Thuwajit, C., 2010. Gene expression profiling of cholangiocarcinoma-derived fibroblast reveals alterations related to tumor progression and indicates periostin as a poor prognostic marker. Mol. Canc. 9, 13. https://doi.org/10.1186/1476-4598-9-13.

Wang, C.A., Liu, Y.C., Du, S.Y., Lin, C.W., Fu, H.W., 2008. *Helicobacter* pylori neutrophil-activating protein promotes myeloperoxidase release from human neutrophils. Biochem. Biophys. Res. Commun. 377, 52–56.

Wehbe, H., Henson, R., Meng, F., Mize-Berge, J., Patel, T., 2006. Interleukin-6 contributes to growth in cholangiocarcinoma cells by aberrant promoter methylation and gene expression. Cancer Res. 66, 10517–10524.

Whitmore, L.C., Weems, M.N., Allen, L.H., 2017. Cutting edge: *Helicobacter pylori* induces nuclear hypersegmentation and subtype differentiation of human neutrophils in vitro. J. Immunol. 198, 1793–1797.

Wilson, M.S., Taylor, M.D., Balic, A., Finney, C.A., Lamb, J.R., Maizels, R.M., 2005. Suppression of allergic airway inflammation by helminth-induced regulatory T cells. J. Exp. Med. 202, 1199–1212.

Wongratanacheewin, S., Rattanasiriwilai, W., Priwan, R., Sirisinha, S., 1987. Immunodepression in hamsters experimentally infected with *Opisthorchis viverrini*. J. Helminthol. 61, 151–156.
Wongratanacheewin, S., Sermswan, R.W., Sirisinha, S., 2003. Immunology and molecular biology of *Opisthorchis viverrini* infection. Acta Trop. 88, 195–207.
Wright, H.L., Moots, R.J., Bucknall, R.C., Edwards, S.W., 2010. Neutrophil function in inflammation and inflammatory diseases. Rheumatology 49, 1618–1631.
Xu, X.Q., Wang, Z.H., Liao, J.X., Chen, X.Y., Liu, W.Z., Xiao, S.D., Lu, H., 2012. Predictive value of neutrophil infiltration as a marker of *Helicobacter pylori* infection. World J. Gastroenterol. 18, 5101–5105.
Xuan, S.Y., Li, N., Qiang, X., Zhou, R.R., Shi, Y.X., Jiang, W.J., 2006. *Helicobacter* infection in hepatocellular carcinoma tissue. World J. Gastroenterol. 12, 2335–2340.
Yeoh, K.W., Promthet, S., Sithithaworn, P., Kamsa-Ard, S., Parkin, D.M., 2015. Re-examination of *Opisthorchis viverrini* infection in Northeast Thailand. Asian Pac. J. Cancer Prev. APJCP 16, 3413–3418.
Yongvanit, P., Pinlaor, S., Bartsch, H., 2012a. Oxidative and nitrative DNA damage: key events in opisthorchiasis-induced carcinogenesis. Parasitol. Int. 61, 130–135.
Yongvanit, P., Thanan, R., Pinlaor, S., Sithithaworn, P., Loilome, W., Namwat, N., Techasen, A., Dechakhamphu, S., 2012b. Increased expression of TLR-2, COX-2, and SOD-2 genes in the peripheral blood leukocytes of opisthorchiasis patients induced by *Opisthorchis viverrini* antigen. Parasitol. Res. 110, 1969–1977.
Yothaisong, S., Thanee, M., Namwat, N., Yongvanit, P., Boonmars, T., Puapairoj, A., Loilome, W., 2014. *Opisthorchis viverrini* infection activates the PI3K/AKT/PTEN and Wnt/beta-catenin signaling pathways in a cholangiocarcinogenesis model. Asian Pac. J. Cancer Prev. APJCP 15, 10463–10468.
Zhao, Y., Yokota, K., Ayada, K., Yamamoto, Y., Okada, T., Shen, L., Oguma, K., 2007. *Helicobacter pylori* heat-shock protein 60 induces interleukin-8 via a Toll-like receptor (TLR)2 and mitogen-activated protein (MAP) kinase pathway in human monocytes. J. Med. Microbiol. 56, 154–164.

FURTHER READING

Glaser, S.S., Gaudio, E., Miller, T., Alvaro, D., Alpini, G., 2009. Cholangiocyte proliferation and liver fibrosis. Expet Rev. Mol. Med. 11, e7.
Songserm, N., Prasongwattana, J., Sithithaworn, P., Sripa, B., Pipitkool, V., 2009. Cholangiocarcinoma in experimental hamsters with long-standing *Opisthorchis viverrini* infection. Asian Pac. J. Cancer Prev. APJCP 10, 299–302.

CHAPTER SEVEN

Recent Advances in the Diagnosis and Detection of *Opisthorchis viverrini* Sensu Lato in Human and Intermediate Hosts for Use in Control and Elimination Programs

Weerachai Saijuntha*, Kanyarat Duenngai[§], Sirikachorn Tangkawattana[¶], Trevor N. Petney[||,#,], Ross H. Andrews[||,§§] and Paiboon Sithithaworn[||,1]**

*Walai Rukhavej Botanical Research Institute, Mahasarakham University, Maha Sarakham, Thailand
[§]Department of Public Health, Faculty of Science and Technology, Phetchabun Rajabhat University, Phetchabun, Thailand
[¶]Department of Vetrinary Pathobiology, Faculty of Veterinary Medicine, Khon Kaen University, Khon Kaen, Thailand
[||]Cholangiocarcinoma Screening and Care Program (CASCAP) and Institute of Cholangiocarcinoma, Khon Kaen University, Khon Kaen, Thailand
[#]Institute of Zoology 1: Ecology and Parasitology, Karlsruhe Institute of Technology, Karlsruhe, Germany
**State Museum of Natural History Karlsruhe, Karlsruhe, Germany
[§§]Faculty of Medicine, St Mary's Campus, Imperial College, London, United Kingdom
[1]Corresponding author: E-mail: paib_sit@kku.ac.th

Contents

Advances in Parasitology, Volume 101
ISSN 0065-308X
https://doi.org/10.1016/bs.apar.2018.05.007

Abstract

Opisthorchiasis is a neglected tropical disease, caused by infection with the fish-borne trematode *Opisthorchis viverrini* sensu lato that afflicts more than 10 million people in Southeast Asia, including Thailand, Lao PDR, Vietnam and Cambodia. The disease is characterized by a chronic infection that induces hepatobiliary inflammation, especially periductal fibrosis, which can be detected by ultrasonography. This chronic inflammation eventually leads to cholangiocarcinoma (CCA), a usually fatal bile duct cancer that develops in approximately 1% of *O. viverrini*–infected individuals. In Thailand alone, CCA kills up to 20,000 people every year and is therefore of substantial public health importance. Its socioeconomic impacts on impoverished families and communities are considerable. To reduce *O. viverrini*–associated morbidity and CCA, the primary intervention measures focus on opisthorchiasis control and elimination. Accurate diagnoses of *O. viverrini* infection, in both mammalian, snail and fish intermediate hosts, are important for achieving these goals. Despite extensive efforts over several decades to find sensitive and specific diagnostics for opisthorchiasis, a simple and robust diagnostic method is still required. Here we review earlier and current developments in the search for new diagnostics for opisthorchiasis, with practical applications in the research laboratory, the clinic and the field. Of the methods currently available, the urine antigen assay shows considerable potential for the diagnosis and screening of opisthorchiasis. Nevertheless, these new assays require validation, determination of their cost-effectiveness when applied for mass screening in an endemic setting in support of policy decisions for national public health programs aimed at the control and elimination of opisthorchiasis.

1. INTRODUCTION

Opisthorchiasis, caused by *Opisthorchis viverrini* sensu lato (*O. viverrini* species complex) infection, is a major public health problem in mainland Southeast Asia particularly Thailand, Lao PDR, Vietnam, Cambodia and Myanmar (Sithithaworn et al., 2012; Aung et al., 2017). Unlike other food-borne trematodiasis, infection with this liver fluke induces hepatobiliary disease and bile duct cancer (cholangiocarcinoma, CCA), and *O. viverrini* was designated as group I carcinogen (IARC, 2012). The extent and impact of *O. viverrini*–associated CCA were not well recognized until recently. In Thailand alone an estimated 20,000+ people die annually of this disease

(Bundhamcharoen et al., 2011; Khuntikeo et al., 2016). To date, there are several known risk factors for CCA, but the most fundamental factor is liver fluke infection (Sithithaworn et al., 2014). To reduce CCA incidence and its resulting fatality, the primary focus has aimed at prevention, control and elimination of the liver fluke in the endemic communities. Under the current transmission landscape of opisthorchiasis, light infections predominate and only a few people carry heavy infections. To screen the light infection group, improved diagnostic methods are required to replace or supplement the traditional methods which have been used for several decades. A robust diagnostic method also serves as an important tool for parasite prevention and control by mass chemotherapy. Currently, there are three major methods in use to diagnose *O. viverrini* infection in humans, as well as in other animal hosts, namely parasitological, immunological and molecular. Furthermore, accurate diagnostic methods are required to detect and confirm *O. viverrini* infection in humans and in animal hosts, including the other live stages in the life cycle such as cercariae in *Bithynia* snails, free-living cercaria in water and metacercariae infecting the second intermediate host, cyprinid fish (Fig. 1). In addition, since only praziquantel is used for chemotherapeutic control, other alternatives are required.

Thus, this chapter aims to review and update procedures for the diagnosis and detection of the various life stages of *O. viverrini*, as well as their implications for epidemiological studies and parasite control.

2. HUMAN DIAGNOSIS

2.1 Conventional Methods

The classical diagnosis of *O. viverrini* infection is recovery of adult worms for morphological identification. Many studies have obtained adult worms by searching for expelled worms in faecal specimens at postanthelmintic treatment (Ramsay et al., 1989; Elkins et al., 1991; Radomyos et al., 1994; Sato et al., 2015). Worm recovery by postanthelmintic treatment or expulsion chemotherapy may not be efficient as unknown proportions of worms were not recovered, and hence there were egg-positive cases who expelled no worms (Elkins et al., 1991). The worm recovery method by expulsion chemotherapy is suitable for species confirmation (Chai et al., 2005; Lovis et al., 2009), but for diagnostic purposes it may be difficult to efficiently handle and process large faecal volumes. By contrast, worm burden can be accurately recovered by collecting the worms directly from

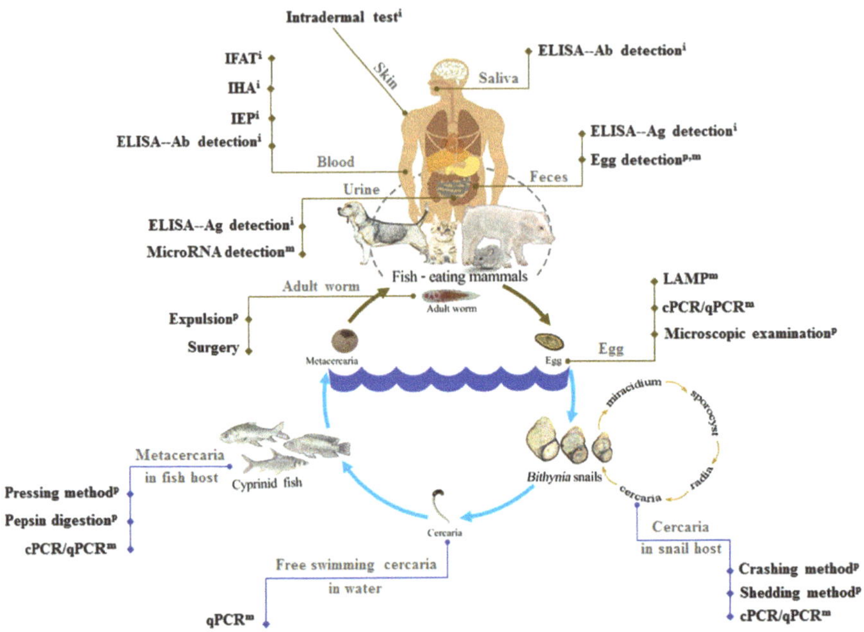

Figure 1 The methods for diagnosis of *Opisthorchis viverrini* infection through its life cycle. p = parasitological method, i = immunological method, m = molecular method. *ELISA*, enzyme-linked immunosorbent assay; *PCR*, polymerase chain reaction.

the liver by postmortem autopsy (Sithithaworn et al., 1991). Based on autopsy study, the detection limit for finding eggs in the faeces using the standard formalin-ethyl acetate concentration technique (FECT) was estimated to require 20 worms or approximately 100 or less EPG. Individuals with low infection intensities and limited egg output are likely to be underdiagnosed by microscopic examination by as much as ~20% (Sithithaworn et al., 1991).

Similar to other intestinal helminthiases, the conventional diagnostic method is egg detection in faeces in human and animal hosts. The detection of eggs in faeces by microscopic examination is considered to be the gold standard for opisthorchiasis diagnosis. In addition, eggs in faecal samples can be detected during treatment of bile duct obstruction either in the bile, from nasobiliary, percutaneous transhepatobilbiary drainage, or in the duodenal fluid (Sithithaworn and Haswell-Elkins, 2003; Saijuntha et al., 2014). Occasionally, eggs can be found in biopsy materials or tissues resected during surgery or at necropsy, which can then be examined directly or can be fixed and examined as stained histological sections for diagnosis and pathological purposes (Riganti et al., 1988, 1989).

However, the egg of *O. viverrini* which can be identified by characteristic rough and thick eggshell is very similar to the minute/small eggs of several species of other food-borne trematodes belonging to the families Opisthorchiidae, Heterophyidae and Lecithodendriidae. The latter two families are commonly referred to as minute intestinal flukes (MIFs) due to the small size of the adult worms compared with the liver flukes (Chai et al., 2005; De et al., 2011). These species are, like the liver flukes, fish-borne trematodes (FBTs) or fish-borne zoonotic trematode (Lan-Anh et al., 2009; Phan et al., 2010a,b). The similarity to the eggs of these FBT species and those of *O. viverrini*, which often coexist in the same community, can substantially increase the likelihood of false positive diagnosis, depending on the prevalence of these species. Thus, apart from adult worm recovery for identification, the accurate diagnosis and differentiation of the minute eggs requires expertise of the microscopists. Parasitological techniques have been progressively developed and improved for easier diagnosis and differentiation of *O. viverrini* eggs from those of other flukes. For instance, fat extraction using diethyl ether or ethyl acetate to eliminate faecal debris from the sediment has made it cleaner for easier microscopic examination, such as in FECT (Elkins et al., 1990). Special staining of faecal specimens or sedimentation with potassium permanganate (Sukontason et al., 1999) or methylene blue (Pasuralertsakul et al., 2005) has been used to highlight the different characteristics of *O. viverrini* eggs from other morphologically similar eggs, such as those of *Haplorchis taichui*, *Phaneropsolus bonnie*, *P. molenkampi* and *Stellantchasmas falcatus*.

Because the collection of faecal samples is not an invasive procedure, detection of eggs in faecal samples is commonly used in routine work for the diagnosis of *O. viverrini* infection. A number of methods are available for faecal examination, including a direct smear, the modified thick Kato smear or Kato-Katz (KK) (Hong et al., 2003; Johansen et al., 2015), the modified FECT (Elkins et al., 1990), Stoll's dilution egg count technique (Viyanant et al., 1983) and recently the commercial faecal concentrator kit (Laoprom et al., 2016a). The KK method has been recommended for the diagnosis of liver fluke infections for both opisthorchiasis and clonorchiasis (infection with *Clonorchis sinensis*). For more accurate diagnosis, a repetitive examination on consecutive 3-day faecal samples by KK or a single examination by FECT was suggested (Sayasone et al., 2015). To circumvent the drawbacks of both KK and FECT, the faecal concentrator kit (mini Parasep, Diasys, UK), which is a closed concentrating system, provides an increased ease of use and presents a reduced health hazard compared with conventional concentration methods. The initial test suggested that it is slightly less sensitive than FECT, but it gave better diagnostic accuracy than the simple smear method for the diagnosis of *O. viverrini* infection and for other parasitic infections (Laoprom et al., 2016a). Although the kit offers a safer and faster way to conduct the concentration procedure and is commonly used in field studies, there are considerable disadvantages in terms of cost and diagnostic accuracy compared with the KK method.

It is known that animal reservoir hosts (cats and dogs) are important in maintaining the *O. viverrini* life cycle and spreading the parasite (Enes et al., 2010; Pumidonming et al., 2016). Thus, the diagnosis and treatment of *O. viverrini* infection in these companion animals are important for the prevention and control of opisthorchiasis in endemic areas. The conventional diagnostic method is again the detection of eggs in faecal samples. It is known that these domestic animals often harbour multiple parasite infections, thus careful identification is required particularly for species that have a similar egg morphology to human parasites. For example, *Opisthorchis tenuicollis* has been recovered in pigs from Mumbai, India (Jawalagatti et al., 2016), *Opisthorchis lobatus* metacercariae were recovered from snakehead fish in Cambodia and can develop to adult worms in hamsters (Thaenkham et al., 2011), *Opisthorchis caninus*, can develop to adult worms in rhesus monkeys (Pande and Shukla, 1974), as well as an *Opisthorchis*-like fluke discovered in ducks (Dao et al., 2014) and other birds in Vietnam. This species was molecularly identified as an *Opisthorchis* sibling species of *O. viverrini* (Dao et al., 2016). Application of more accurate diagnostic methods to assess the true prevalence of infection in these animal reservoir hosts, such as immunological or molecular methods, is not available.

2.2 Immunological Methods

Immunodiagnostic tests have the advantage of being applicable during all stages of the disease, and consequently they are commonly used for the diagnosis of infections in field situations. Several serological tests for human opisthorchiasis have been developed for use in a diagnostic assay with greater sensitivity and specificity than the faecal examination technique. These include the intradermal test, immunoelectrophoresis, indirect haemagglutination assay, indirect fluorescent antibody test and indirect enzyme-linked immunosorbent assay (indirect ELISA) (Wongratanacheewin et al., 2003). For these diagnostic methods, the type(s) of antigen (Ag) and reacting antibody (Ab) play an important role in the sensitivity and specificity of the test. Thus, selection of an appropriate antigen and antibody molecule for immunological diagnosis is a crucial step. There are, however, many candidate molecules that could serve as antigen or induced antibody for the diagnosis of *O. viverrini* infection (Fig. 2).

Immunological parameters have been used as biomarkers for parasite-induced pathologies. Measurements of delayed-type hypersensitivity responses to unrelated antigens, and liver fluke-specific IgG and IgA levels in serum were correlated with the biliary tract abnormalities of residents from an endemic community, as determined by ultrasonography (Haswell-Elkins et al., 1991; Mairiang et al., 1992). Immune responsiveness to unrelated antigens did not vary with the intensity of parasite infection or disease status. Of all the variables, IgG levels were most markedly elevated in disease cases compared with normal subjects and were closely associated with gall bladder size and dysfunction. This is consistent with the hypothesis that an immunopathological mechanism is involved in opisthorchiasis and

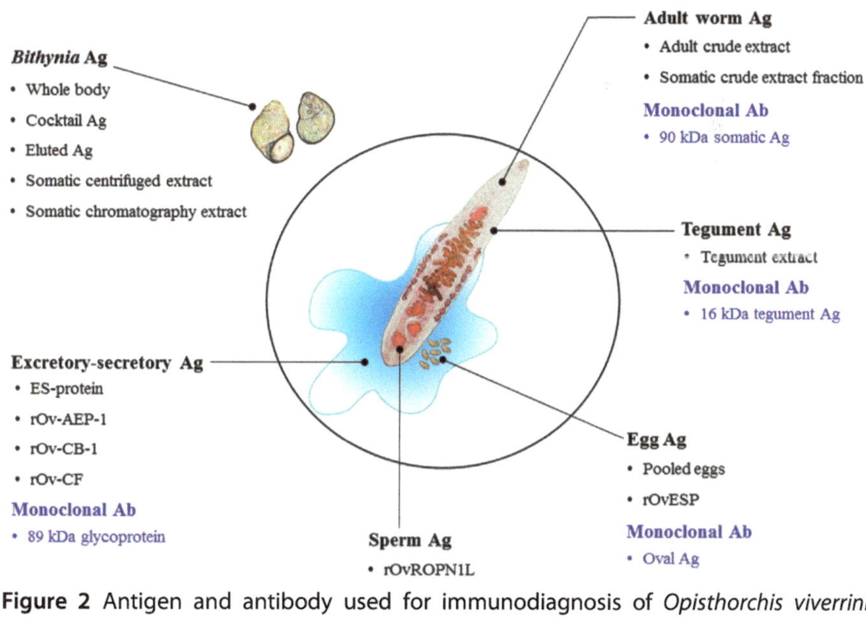

Figure 2 Antigen and antibody used for immunodiagnosis of *Opisthorchis viverrini* infection. *AEP*, asparaginyl endopeptidase; *CB*, cathepsin B; *CF*, cathepsin F; *ES*, excretory–secretory.

suggests that antibody levels may be useful in screening populations for fluke-associated hepatobiliary disease (Haswell-Elkins et al., 1991; Sripa et al., 2012a). The persistency level of parasite-specific IgG in serum has been utilized to indicate past *O. viverrini* infection and risk factor for CCA (Sithithaworn et al., 2014).

On the serodiagnosis front, several types of liver fluke-derived antigens and recombinant proteins were evaluated for diagnostic usefulness. Crude somatic extract of adult *O. viverrini* was the most common antigen used for antibody detection by ELISA (Haswell-Elkins et al., 1992; Wongratanacheewin et al., 2003), providing higher diagnostic performance than faecal examination (Poopyruchpong et al., 1990; Wongsaroj et al., 2001). The composition of somatic extract of adult *O. viverrini* was identified by radioimmunoprecipitation and polyacrylamide gel electrophoresis. It was composed of protein sizes of >116, 89, 78 and 20 kDa. Several proteins in the extract reacted specifically to human opisthorchiasis serum. A protein with molecular weight 89 kDa was found the most specific antigen for the diagnosis of *O. viverrini* infections in humans (Wongratanacheewin and Sirisinha, 1986). This antigenic protein was dominant in the excretory—secretory proteins (ES product) from adult worms (Wongratanacheewin et al., 1988).

An extraction of surface tegument proteins also reacted specifically to sera of human opisthorchiasis, i.e., >116, 108, 64, 38, 34, 20 and 16—17 kDa proteins. Of these, the 16—17 kDa protein revealed the highest concentration, but low immunogenicity compared with 89 kDa protein (Sirisinha et al., 1990). Production of these parasite-derived antigens, especially the ES antigen, is costly because it requires both a laboratory and an experimental animal facility. As such, the supply of antigens for research and commercial exploitation is limited.

With the advances in liver fluke genomics and proteomics, several recombinant antigenic molecules have been examined and assessed for their diagnostic potential. The recombinant antigen produced from eggs and the glycine tyrosine—rich eggshell proteins (rOvESP) of *O. viverrini* were used for serum antibody detection by ELISA (Wongsaroj et al., 2001; Ruangsittichai et al., 2006). Proteins in the ES product examined and synthetic peptide antigen produced were asparaginyl endopeptidase (rOv-AEP-1), cathepsin B-1 protease (rOv-CB-1) and cathepsin F (rOv-CF) (Laha et al., 2008; Sripa et al., 2012b; Teimoori et al., 2015), as well as the rhophilin-associated tail protein of sperm (rOvROPN1L) (Rattanachan et al., 2014). In addition, the propeptide of cathepsin F and cathepsin L-like has been characterized and showed better diagnostic sensitivity and specificity to the conventional faecal examination methods (Kaewpitoon et al., 2008; Pinlaor et al., 2009). As more recombinant proteins are produced, such as of the *C. sinensis* proteins (Li et al., 2011), a suitable antigen(s) for better diagnosis and screening of the high risk human population should be forthcoming in the near future.

As a snail-borne trematode, *O. viverrini* has an intimate evolutionary relationship with its *Bithynia* snail intermediate hosts (Saijuntha et al., 2007). Most interestingly, there is a common antigenic protein in the parasite and its snail host: antigen extracted from the *Bithynia* snails can be used to detect a specific antibody for diagnosis of human *O. viverrini* infection. The association between *O. viverrini* and its snail intermediate hosts, *Bithynia funiculata*, *Bithynia siamensis siamensis* and *Bithynia siamensis goniomphalos*, was determined based on the correlation between shared antigens and infection rates. Crude extract of *O. viverrini* adult worms produced the same number of two common precipitin bands of four different electrophoretic mobilities against anti-*B. funiculata* and anti-*B. s. siamensis* sera and one precipitin band in the antigen well region against anti-*B. s. goniomphalos* serum. The three snail antigens produced the same number of two common precipitin bands against anti-*O. viverrini* serum (Chanawong et al., 1990). Four batches of crude antigens, i.e., adult *O. viverrini* including whole body, head—foot and visceral mass of *B. funiculata* were analyzed by IgG-ELISA. *B. funiculata* antigen was the most effective for the diagnosis of *O. viverrini* infection. However, cross-reactions may occur with sera from patients with paragonimiasis and strongyloidiasis (Watthanakulpanich et al., 1997). Cocktail antigen was prepared using Sephacryl S-200 HR gel filtration chromatography with electroeluted *Bithynia* snail antigens being found at 53 kDa from polyacrylamide gel. The eluted antigens have a higher sensitivity and specificity for diagnosis using indirect ELISA. However, cross reactions with the cocktail antigen were seen in some other infections such as hymenolepsiasis and strongyloidiasis (Waikagul et al., 2002). In a subsequent study, total IgG and IgG1-4 were selectively reactive to antigens of *B. s. goniomphalos* extract derived from liquid-phase isoelectric focusing (IFE). Antigens (Iso-AgF) from 20 IEF fractions were selected based on a high ELISA-OD ratio between pooled-positive and pooled-negative sera. Iso-AgF 7, 7, 6, 2 and 10 resulted in high OD-ratios to total IgG, IgG1, 2, 3 and 4, respectively. Iso-AgF7 to IgG1 showed the best result, with sensitivity, specificity, positive and negative predictive value of 100%, 96%, 86% and 100%, respectively. Low cross-reactivity of IgG1 was found in one case each of gnathostomiasis, trichinellosis, toxocariasis, angiostrongyliasis, bancroftian filariasis, enterobiasis, neurocysticercosis and taeniasis (Pakdee et al., 2010).

Taken together, the *Bithynia* snail antigens approach here may become a promising alternative to the parasite-derived antigens. It is, however, important that any subsequent research should be based within the molecular/genetic framework of the discovery that *B. s. goniomphalos* is a species complex (Saijuntha et al., 2007; Laoprom et al., 2012; Kiatsopit et al., 2014) with may cryptic species associated with cryptic species in the *O. viverrini* species complex in defined wetland systems.

Blood collection by venipuncture requires either a phlebotomist or trained nurse and is an invasive method of sample collection. Alternatively, clinical samples such as urine and saliva are easier to collect; this is done noninvasively and is better accepted in the human population. Antibody detection by ELISA in nonblood samples, such as urine and saliva, has received considerable attention. ELISA used to detect IgG and IgG4 in the urine of opisthorchiasis patients revealed that IgG detection is a promising diagnostic method for opisthorchiasis (Tesana et al., 2007). Moreover, the detection of urine IgG specific to *O. viverrini* antigen may indicate a degree of renal pathology in opisthorchiasis patients, in addition to a greater risk of developing cholangicarcinoma (Saichua et al., 2013). In another study, the patient's saliva was found to have more potential for use in antibody detection than urine (Sawangsoda et al., 2012), and the level of saliva antibody was correlated with an active opisthorchiasis (Chaiyarit et al., 2011).

As opposed to antibody detection, antigen detection is one of the attractive alternatives in the diagnosis of opisthorchiasis. Antigen detection reflects the current status of infection and is relevant for treatment and control. Several reports have tested the usefulness of determining the presence of ES products of *O. viverrini* in faeces. An 89 kDa antigen in the ES products has been found to be a promising candidate for coproantigen detection by ELISA (Amornpunt et al., 1991; Sirisinha et al., 1991). A monoclonal antibody—based enzyme-linked immunosorbent assay (MAb-ELISA) has been developed to detect antigen for the diagnosis of ongoing opisthorchiasis. Specific monoclonal antibodies have been produced against *O. viverrini* antigenic peptides such as 16 kDa tegument protein, 90 kDa somatic protein and 89 kDa glycoprotein (Amornpunt et al., 1991). The sensitivity was higher than microscopic detection of *O. viverrini* eggs (Sirisinha et al., 1995). Another study indicated sensitivity 31% and specificity of 100%

(Chaicumpa et al., 1992). On the other hand, Sirisinha et al. (1991) reported a sensitivity and specificity using MAb-ELISA of 57.1% and 70.5%, respectively, using a monoclonal antibody reactive to the 89 kDa ES antigen of *O. viverrini*. Treatment of faecal samples with modified trichloroacetic acid improved detection sensitivity of MAb-ELISA to 52 ng/mL and provided a quantitative diagnosis (Watwiengkam et al., 2013). Recently, coproantigen detection by sandwich ELISA using recombinant *O. viverrini* cathepsin F chicken IgY in combination with rabbit IgG antibody to the somatic *O. viverrini* antigens was investigated. This method showed a detection limit of 70 ng/mL by spiking the parasite antigens into control faeces. When applied for the diagnosis, the IgY-based sandwich ELISA exhibited sensitivity and specificity of 93.3% and 76.7%, respectively (Teimoori et al., 2017). To change from faecal to urine samples, recently, a novel urinary antigen detection method was established for the quantitative diagnosis of opisthorchiasis by MAb-ELISA with 81% sensitivity and 70% specificity when compared with the current gold standard diagnostic method (Worasith et al., 2015). The advantage of using urine samples is the noninvasiveness of sample collection and the high acceptability by individuals. These antigen detection methods offer many advantages, being simple, reproducible and allowing for the simultaneous analysis of many samples without any special equipment. This technique may be used for mass screening, especially in new endemic areas. It also detects only current infections.

2.3 Molecular Methods

There are, nevertheless, limitations to the parasitological and immunological methods of diagnosis, such as the difficulty of differentiating the eggs of *O. viverrini* from those of other fish-borne trematodes, the detection of light infections, the limited immune response period and cross-reactions in immunodiagnosis. Molecular diagnosis represents an alternative choice for solving these problems. A number of target genes from *O. viverrini* have been tested for their diagnostic suitability, including satellite DNA, ITS1, ITS2, 28S rRNA, mitochondrial DNA and microRNA. Several genetic markers/techniques involving conventional polymerase chain reaction (PCR) alone or combined with other techniques, e.g., PCR-restriction fragment length polymorphism (RFLP), multiplex PCR, pyrosequencing and LAMP and real-time PCR, can potentially be used to differentiate *O. viverrini* from other species (Table 1). In addition, molecular identification techniques can be used in cases of multiparasite infections in a single host (Sato et al., 2009; Thaenkham et al., 2011). The molecular methods

Table 1 Molecular Diagnosis of *Opisthorchis viverrini* Infection

Molecular Method	DNA Target	Specimens	Purpose of Study	References
cPCR	pOV-A6 DNA	Egg spite in hamster feces	Diagnosis/detection	Wongratanachewin et al. (2001)
cPCR	pOV-A6 DNA	Feces	Diagnosis/detection	Stensvold et al. (2006) and Duenngai et al. (2008)
cPCR	pOVCRA26-5 DNA	Fish tissue/ metacercariae	Diagnosis/detection	Parvathi et al. (2008)
cPCR	pOVCRA26-5 DNA	Feces	Diagnosis/detection	Umesha et al. (2008)
cPCR	pOV-A6 DNA	Snail and fish tissue	Diagnosis/detection	Maleewong et al. (2003)
mPCR	Mitochondrial DNA	Adult, metacercariae and egg	Differential diagnosis	Le et al. (2006)
cPCR combined qPCR	CO1	Feces	Detection	Lamaningao et al. (2017)
qPCR	CO1	Feces	Diagnosis/detection	Hashizume et al. (2017)
mqPCR and HRMA	ND2	Feces	Differential diagnosis	Kaewkong et al. (2013)
TaqMan-qPCR	RPLO gene	Liver tissue	Detection	Suksumek et al. (2008)
FRET-PCR	pOv-A6 DNA	Feces	Detection	Intapan et al. (2009)
FRET-PCR and melting curve analysis	pOv-A6 DNA	Infected fish	Detection	Intapan et al. (2008)
FRET-PCR and melting curve analysis	pOV-A6 DNA	Feces	Differential diagnosis	Janwan et al. (2011)
FRET-PCR	pOV-A6 DNA	Infected Bithynia snails	Detection	Sri-Aroon et al. (2011)
FRET-PCR and melting curve analysis	ND2	Feces	Differential diagnosis	Sanpool et al. (2012)
qPCR and HRMA	CO1	Adult DNA	Differential diagnosis	Cai et al. (2014)
cPCR and pyrosequencing	28S rRNA	Adult DNA and feces	Differential diagnosis	Tantrawatpan et al. (2014)
cPCR, PCR-RFLP and DNA sequencing	18S-ITS1-5.8S rDNA	Adult DNA	Differential diagnosis	Kang et al. (2007)

cPCR and DNA sequencing	CO1	Feces	Confirmation identification	Aung et al. (2017)
cPCR and DNA sequencing	ITS2	Cercaria from infected snails	Confirmation identification	Boonmekam et al. (2017)
cPCR, DNA sequencing and qPCR	CO1	Feces	Differential diagnosis	Lamaningao et al. (2017)
cPCR and DNA sequencing	CO1	Adult DNA	Confirmation identification	Vo et al. (2014)
cPCR and DNA sequencing	CO1 and ND1	Feces	Diagnosis and identification	Buathong et al. (2015, 2017)
cPCR, PCR-RFLP and DNA sequencing	ITS2	Feces	Diagnosis/identification	Traub et al. (2009) and Buathong et al. (2017)
Loop-mediated isothermal amplification	ND1	Feces	Diagnosis	Le at al. (2012)
	ITS1	Feces	Diagnosis	Arimatsu et al. (2012, 2015)
	Microsatellites (OVMS6)	Feces	Diagnosis	Arimatsu et al. (2015)
qPCR	MicroRNA	Urine	Diagnosis	Silakit et al. (2017)
cPCR	Retrotransposon DNA (OV-RTE-1)	Adult DNA and feces	Diagnosis	Thi Phung et al. (2014)
qPCR	CO1	Water	Detection in environment	Hashizume et al. (2017)
Multiplex ligation-dependent probe amplification	ITS1	Adult DNA	Differential diagnosis	Sun et al. (2011)
HAT-RAPD	Unknown	Adult DNA	Differentiation	Wongsawad and Wongsawad (2012)

dqPCR, duplex real-time PCR; *FRET*, fluorescence resonance energy transfer; *HAT-RAPD*, high annealing temperature-random amplified polymorphic DNA; *mqPCR*, multiplex real-time PCR; *PCR*, polymerase chain reaction; *RFLP*, restriction fragment length polymorphism.

discussed below will contribute significantly towards a more effective and accurate diagnosis of trematode infections, although further simplification of the tests and an understanding of cost-effectiveness under various socioeconomic scenarios are needed. In addition, the validation of DNA positive test results is required, although evidence from animal models is accumulating and is supported by human studies (Rahman et al., 2011; Duenngai et al., 2013). Such approaches can also be used in a food security assessment to test for the presence of liver flukes in aquaculture or native fisheries products, particularly for export (Parvathi et al., 2007, 2008; Cai et al., 2010). Due to their high specificity, such molecular diagnostic tests are likely to play an increasing role in anthelminthic drug efficacy evaluations, the rigorous monitoring of reinfection patterns, and to investigate changes in the endemic range of the liver flukes (Touch et al., 2009; Chen et al., 2010).

The early period of the molecular diagnosis of *O. viverrini* infection focused on developing species-specific primers for detecting *O. viverrini* DNA. The repeat DNA element (pOV-A6 DNA) was shown to be a good molecular marker for this purpose. Primer OV-6 was proven to be specific for pOV-A6 DNA of *O. viverrini* with high sensitivity and specificity (Wongratancheewin et al., 2001). The detection of *O. viverrini* egg DNA in human stools using PCR and based on primers complementary to the repeat DNA element showed a specificity of 98% and a sensitivity of 100% for moderate to heavy infections with more than 1000 EPG. In light infections with less than 200 EPG, the sensitivity was reduced to only 68% (Wongratancheewin et al., 2001, 2002). In another, subsequent PCR-based study using the same target DNA, pOV-A6 showed low sensitivity (50%) at high egg counts of more than 1000 EPG in faecal samples from Lao PDR (Stensvold et al., 2006). However, if the quality of the DNA was improved by using cetyltrimethylammonium bromide during its preparation (CTAB) to remove PCR inhibitors, the sensitivity was increased to 79% (Duenngai et al., 2008). The pOVCRA26-5 DNA method was subsequently developed and used for specific detection of *O. viverrini* DNA in human faeces with high sensitivity and specificity (Umesha et al., 2008). More recently, the retrotransposon of *O. viverrini* (*OV-RTE-1*) was found to be a new alternative genetic marker of high sensitivity and specificity for the PCR diagnosis of opisthorchiasis. PCR-positive tests occurred in 29% of cases that were parasite-negative using the conventional faecal examination method, indicating its potential diagnostic value for light infections (Thi Phung et al., 2014). PCR methods targeting the CO1 and ND1 genes to detect *O. viverrini* in faecal specimens have been developed. The sensitivities of PCR for CO1 and ND1 were 89.1% and 71.7%, respectively, in specimens containing greater than 100 EPG *O. viverrini* eggs (Lovis et al., 2009; Buathong et al., 2015).

Conventional PCR combined with other techniques such as RFLP (PCR-RFLP) of 18S-ITS1-5.8S rDNA can be used to reliably differentiate between *O. viverrini, C. sinensis* and *Opisthorchis felineus* infections (Kang et al., 2008). PCR-RFLP of the ITS2 region can be used to differentiate *O. viverrini* from *C. sinensis*. This method was used to detect *C. sinensis* DNA from Central Thailand (Traub et al., 2009). Subsequently, use of PCR-RFLP of the ITS2 region and DNA sequencing of CO1 and ND1 genes for confirmation of the identification of *O. viverrini*-like eggs detected in faecal samples of the villagers from Chachoengsao Province, Central Thailand found no *C. sinensis* in this endemic *O. viverrini* area (Buathong et al., 2017), as previously reported by Traub et al. (2009). The HAT-RAPD PCR method was used to differentiate *O. viverrini* from *H. taichui*. Two pairs of specific primers were designed to amplify *H. taichui* and *O. viverrini*. These generated a specific PCR products of 170 and 319 bp, respectively, with no cross-reaction with any of the other parasites tested (Wongsawad and Wongsawad, 2012). A combination of cPCR and real-time PCR (qPCR) of the CO1 region for detecting and discriminating between *O. viverrini* and *H. taichui* infections was developed for testing clinical stool samples and exhibited a sensitivity of 100% and specificity for *O. viverrini* and *H. taichui* of 71% and 91%, respectively (Lamaningao et al., 2017).

Another PCR-based molecular identification method is multiplex ligation—dependent probe amplification using the ITS1 region, which allows rapid and specific detection of single nucleotide acid differences between *O. viverrini, C. sinensis* and *O. felineus* (Sun et al., 2011). A PCR assay coupled with DNA pyrosequencing was developed for the identification of FBTs, including *O. viverrini*. A region of 28S rRNA with a length of 46—47 nucleotides contained variable positions for species-specific identification that were selected for primer design and PCR amplification. This method can differentiate *O. viverrini* from the other FBTs, i.e., *C. sinensis, H. taichui, H. pumilio* and *S. falcatus* by seven nucleotide positions (Tantrawatpan et al., 2014). The multiplex approach is potentially useful not only for human diagnosis but also for fisheries product inspection.

Loop-mediated isothermal amplification (LAMP) is another potential molecular method that is widely used for the detection and identification of parasitic trematodes. It has been established for the detection of *O. viverrini* with a higher sensitivity than microscopic examination and conventional PCR. The mitochondrial ND1 region was used by LAMP to detect *O. viverrini* DNA (mito-OvLAMP) using four LAMP primers. The specificity was 100% when tested with *C. sinensis*, *H. pumilio*, *H. taichui*, *Fasciola hepatica* and *F. gigantica*. The limit of DNA detection was very low at 100 fg, indicating 10 to 100 times more sensitivity than cPCR (Le et al., 2012). Although the ITS1 region was selected for LAMP analysis (ITS1-LAMP) with five primers this revealed high sensitivity but low specificity because other parasites were also amplified (Arimatsu et al., 2012, 2015). Thus, the other species-specific region, namely microsatellite 6 (OVMS6), was selected for LAMP analysis (HNB-LAMP). This showed a specifically amplified region of DNA from *O. viverrini*, but not from *C. sinensis*, *O. felineus*, *Centrocestus caninus*, *H. taichui*, *F. gigantica* or *Haplorchoides* sp. This had sensitivity 1000 times higher than ITS1-LAMP, which only needed 1 ng of DNA to be detected (Arimatsu et al., 2015).

Recently, real-time or quantitative PCR (rt-PCR or qPCR) has been widely used for the diagnosis of human parasitic infections. TaqMan real-time PCR assay was used to detect *O. viverrini* DNA in the liver tissue of hepatocellular carcinoma (HCC) and CCA patients (Suksumek et al., 2008). Real-time fluorescence resonance energy transfer (FRET) PCR was consequently developed to detect the pOV-A6 DNA of *O. viverrini* in human faecal samples with high sensitivity and specificity (Intapan et al., 2009; Janwan et al., 2011), as well as to differentiate *O. viverrini* and *C. sinensis* eggs in human faecal samples by real-time FRET PCR with a melting curve analysis of the ND2 region (Sanpool et al., 2012). Multiplex real-time PCR coupled with high-resolution melting (HRM) analysis of the ND2 region for the differentiation of *O. viverrini* and *C. sinensis* eggs in faecal samples has been developed (Kaewkong et al., 2013). An integrating assay of real-time PCR and HRM analysis of the CO1 region for the specific detection and rapid identification of *C. sinensis* and *O. viverrini* has also been developed (Cai et al., 2014). A new real-time PCR of the CO1 region has been developed for the detection of *O. viverrini* DNA in human faecal and environmental water samples in Lao PDR (Hashizume et al., 2017).

The availability of highly specific molecular methods has led to the idea of multiple parasite detection that can give a complete transmission picture of NTDs allowing more suitable and better prevention and control (Llewellyn et al., 2016; Meurs et al., 2017).

3. DETECTION FROM THE INTERMEDIATE HOSTS AND FROM THE ENVIRONMENT

To complete its life cycle, *O. viverrini* needs two intermediate hosts, i.e., the first is *Bithynia* snails and the second is freshwater cyprinid fish. Knowledge of the prevalence of infection in these intermediate hosts is essential to confirm parasite endemicity and transmission dynamics. For transmission, the developed stages need to pass through environment between hosts to reach the new hosts, i.e., eggs have to enter the freshwater and wait to be eaten by *Bithynia* snails. In these snails, they develop into cercariae that are shed into the water as free-swimming cercariae and seek a cyprinid fish in which they become metacercariae. Thus, our ability to detect the various stages *O. viverrini* within the species complex is critical for our understanding of the eco-epidemiology of this parasite.

3.1 Snails

The intramolluscan developmental stages of *O. viverrini* in *Bithynia* snails consist of miracidium, sporocyst, redia and cercaria. Of these stages, the cercariae are the most easily detected as they shed into the water to find their next host (Kiatsopit et al., 2012). Usually, snails are individually placed in small containers and incubated for some time under constant illumination and temperature to induce shedding. After this, the emerged cercariae can be identified under a stereomicroscope and their prevalence recorded. Cercariae can be stained by 0.5% neutral red and identified according to their specific characteristics as pleurolophocercus cercaria (formed of a body and tail), the body has two suckers, a primitive gut, two eye spots, five pairs of penetration glands, cystogenous glands and an excretory bladder. The tail is simple and longer than the body and is provided with cubicular sheath. The cercariae do not emerge during darkness, but low-intensity light could induce a release (Phongsasakulchoti et al., 2005). The seasonal pattern of *O. viverrini* s.l. cercarial emergence reveals that the hot-dry season shows the highest prevalence, which declines in the rainy and cool-dry seasons. Peak cercarial emergence occurred between 08:00 and 10:00 h during the rainy and cool-dry seasons and between 10:00 and 12:00 h during the hot-dry season (Kiatsopit et al., 2012; Laoprom et al., 2016b). Light intensity is the most important stimulus determining the quantity of *O. viverrini* cercariae shed from naturally infected *Bithynia* snails. Snails were evaluated for cercariae output every hour after exposure to various light intensities for a total period of 7 h. The same infected snail was tested under different intensities of light: in the dark, and at 1000, 3000 and 5000 lx. The data showed that under exposure to 1000 and 3000 lx of light, the average percentage and number of cercariae released were higher than when exposed to 5000 lx during the first 2 h of the experiment. In contrast, under higher illumination (5000 lx) a longer time (6 h) was required to stimulate the peak emergence of cercariae. Among the three intensities of light, exposure at 1000 lx induced the highest average number of released cercariae per snail and the highest percentage of cercarial emergence within the first 2 h (54.86%), followed by exposure at 3000 lx (25.58%) and 5000 lx (7.78%). The results suggest that the light intensity of 1000 lx for 2 h would be optimal for *O. viverrini* cercarial shedding from naturally infected *B. s. goniomphalos* snails (Kaewkes et al., 2012).

To increase the accuracy of nondestructive method and also to detect prepatent infections with trematode larvae, multiple sequential shedding for several days or weeks can be used (Studer and Poulin, 2012). Other methods are rarely used as they are time-consuming, for example, enzymatic electrophoresis of the snail's digestive glands, which allows the detection and identification of immature stage infection (Caron et al., 2008). Trematode cercarial release as a detection method has been criticized as inaccurately by several studies (Born-Torrijos et al., 2014). A very low prevalence of *O. viverrini* infection in *Bithynia* snails (estimated lower than 0.01%) has been reported using this detection method in several endemic areas (Ngern-klun et al., 2006). Moreover, in some cases snails containing mature cercariae did not shed (Stunkard and Hinchliffe, 1952). However, there are reports of varying prevalence of *O. viverrini* infection in *Bithynia* snails in some endemic areas in Thailand, Lao PDR and Vietnam (Kiatsopit et al., 2012; Dao et al., 2017), which could reflect that we are dealing with cryptic species of both that may have coevolved and may have differing biological properties/responses etc. In Vietnam, *O. viverrini* cercariae were observed in *B. s. goniomphalos* and *B. funiculata*, at infection rates of 0.86% and 0.14%, respectively (Dao et al., 2017).

Investigating infection levels in snails has already highlighted the importance of PCR-based techniques (Born-Torrijos et al., 2012). The real-time FRET PCR method of pOV-A6 DNA revealed an *O. viverrini* infection rate in snails of 8.52%, whereas using the cercarial shedding method this was 2.47% (Sri-Aroon et al., 2011). The conventional PCR of pOV-A6 showed a high sensitivity and specificity to *O. viverrini* cercarial DNA in infected *Bithynia* snails (Maleewong et al., 2003). Moreover, based on currently available DNA sequences of *O. viverrini* submitted in GenBank, confirmation of the identification of *O. viverrini* becomes easier using blast DNA sequences. For example, the ITS2 DNA sequence was used for confirmation of the identification of *O. viverrini* cercariae shed by *Bithynia* from Cambodia (Boonmekam et al., 2017). Moreover, there are several molecular markers/methods that specifically detect *O. viverrini* DNA, such as LAMP, but this has not yet been applied to cercarial DNA in infected *Bithynia*. Thus, future research may require the application of modern molecular techniques, e.g., next generation sequencing for *O. viverrini* DNA detection in *Bithynia*.

3.2 Fish

The infection rate of *O. viverrini* in fish is usually higher than in the first intermediate hosts, *Bithynia* snails (Sithithaworn and Haswell-Elkins, 2003; Petney et al., 2012). Detection of *O. viverrini* metacercariae in their fish hosts is based on morphological identification (Chai et al., 2014). The method of recovering metacercariae from fish is usually done by using the pepsin digestion method (Sithithaworn et al., 1997; Srisawangwong et al., 1997). The sediment contains the metacercariae which can be identified under a stereomicroscope. In a study on cyprinid fish of 13 species collected from 20 provinces in northeastern Thailand, *O. viverrini* metacercariae were found in six species, *Cyclocheilichthys armatus*, *Systomus rubripinnis* (=*Puntius orphoides*), *Hampala dispar*, *Gymnostomus siamensis*, *Osteochilus vittatus* (=*Osteochilus hasselti*) and *Puntioplites proctozysron* (fish taxonomy based on Kottelat, 2013). Among the sites where *O. viverrini* metacercariae–infected fish were found, 70.0% were dams, 23.7% were ponds/lakes and 7.7% were rivers (Pinlaor et al., 2013). *O. viverrini*–like metacercariae were found in 10 fish species representing both Cyprinidae and noncyprinid families. The prevalence of *O. viverrini* infection in fish was significantly associated with species. *Carassius auratus*, a fish species commonly eaten raw, *Rasbora aurotaenia* and *Puntius brevis* had the highest prevalences of 74.0%, 55.8% and 31.6%, respectively. Sharing of the same snail and fish intermediate host species was found for *O. viverrini* and an *O. viverrini* duck genotype that are sympatric in central Vietnam (Dao et al., 2017). A total of 1479 freshwater cyprinid fish comprising 20 different species was surveyed for *O. viverrini* metacercariae in southern Cambodia. *O. viverrini*–metacercariae were found in 10 species with the infection rate ranging 2.1%–66.7%. Adult worms were obtained by experimental infection in hamsters. The liver flukes found in this study were identified as *O. viverrini* by the morphological features of adult worms, and this identification was confirmed by partial CO1 sequencing of the metacercariae (Touch et al., 2009).

Because the morphology of fish-borne trematode metacercariae, i.e., *O. viverrini* and MIFs such as *H. taichui*, are similar, molecular confirmation is sometimes required. The PCR procedure for the detection of *O. viverrini* in natural and experimentally infected cyprinid fish was developed by Pitaksakulrat et al. (2013). This procedure was based on primers designed from pOV-A6 DNA (Wongratanacheewin et al., 2001). The detection was accomplished in host tissue homogenates to which a single metacercaria was introduced. PCR can detect from as little as a single metacercaria in a fish sample. The method gave 100% detectability for all samples (Maleewong et al., 2003). This assay is specific, with no cross-reactions with other trematodes commonly found in fish, including the closely related species *C. sinensis*. The sensitivity of the assay was determined 10^{-12} ng of *O. viverrini* DNA, whereas in artificially spiked fish meat, three metacercariae could be detected (Parvathi et al., 2008). The real-time FRET PCR of pOV-A6 DNA could also detect as little as a single metacercaria artificially inoculated in fish samples (Intapan et al., 2008).

3.3 Environment

The life cycle stages of *O. viverrini* need a period in the external environment, i.e., soil and water, for transmission between hosts of the adult worms who produce eggs and of cercariae. Thus, eggs and cercariae should be detectable in these. Due to the small size of the eggs and cercariae, to directly observe them is extremely difficult, thus the most efficient method is DNA detection using molecular methods. This so-called environmental DNA (eDNA) indicates the presence of organisms contaminating in environmental water or soil samples (Ficetola et al., 2008; Minamoto et al., 2012). Currently there has only been limited work on the detection of *O. viverrini* DNA in the environment. A species-specific real-time PCR assay using the CO1 region for detecting *O. viverrini* using eDNA was developed and the efficacy of eDNA analysis and its applicability in the field were tested using a total of 94 environmental water samples collected from 44 sites in Savannakhet, Lao PDR during May and October 2015 and February 2016. *O. viverrini* eDNA was detected in five samples by real-time PCR, indicating the presence of the fluke in the area and the risk of infection for individuals consuming fish from these water sources (Hashizume et al., 2017). It remains to be seen whether such an application is related to transmission in endemic communities.

4. IMPLICATIONS FOR EPIDEMIOLOGY, PREVENTION AND CONTROL

Accurate and reliable diagnostic methods are essential for epidemiological studies and play an important role in developing prevention and control strategies for diseases at the community and clinical levels. The diagnostic standard for active opisthorchiasis is the presence of viable eggs in faeces. Several conventional methods have been used such as Stoll's dilution egg count, FECT and KK. While quantitative FECT has been preferred because of the sample storage time after being fixed in formalin (Elkins et al., 1990), KK has been used extensively for cross-sectional surveys and monitoring drug efficacies (Sayasone et al., 2015, 2016, 2017, 2018) due to simplicity of the techniques. All of these conventional microscopy—based methods have a common weak point as they are operator dependent and hence diagnostic accuracy depends largely on the experience of the microscopist and to a lesser extent on sample preparation. To circumvent the low sensitivity of the faecal examination method, repeated sample examinations over 3 days or a single FECT has been recommended (Sayasone et al., 2015). FECT analyzes a greater specimen sample of up to 2 g of faeces but requires laboratory equipment that is not always available in most field settings. KK, on the other hand, uses a smaller sample specimen (approx. 41 mg) but requires no special equipment except the regents and the microscope. The all in one tube faecal concentrator kit (with fixative and built in filter system) offers an intermediate between FECT and KK using approximately 0.5 g of faecal material. Field trials for screening for opisthorchiasis using the faecal concentrator kit (mini Parasep, Diasys, UK) were initiated and the kit provided results comparable to that of FECT (Laoprom et al., 2016a). A similar comparison with KK was evaluated and the kit gave a comparable diagnostic performance (Kopolrat, unpublished data). Although a more in-depth study is needed, the fact that the faecal concentrator kit can be kept over several months after being suspended and fixed in formalin similar to FECT means that it offers some advantages over the KK method. From the public health perspective, these conventional faecal examination methods require experience microscopist and also an adequate laboratory facility. These methods are thus far from ideal for field work in a resource poor setting so that a simplified method is needed to facilitate and enhance control efforts.

The Thai Ministry of Public Health (MOPH) launched a national policy to control opisthorchiasis and CCA to be operated from 2015 to 2026. In 2015, the targeted population screened by faecal examination (Kato-thick smear) involved 76,000 people in 84 subdistricts in 27 provinces in the north, northeast and central Thailand. The urine assay for antigen detection based from our previous work (Worasith et al., 2015) was applied in 23 subdistricts in addition to faecal examination (n > 20,000). This initial survey at base line showed that the average prevalence of opisthorchiasis was 11% by faecal examination (combining Kato thick smear and FECT), but this proved to be 41% by urine antigen assay (Sithithaworn, unpublished data). The availability of a sensitive urine assay has, for the first time, allowed us to map the prevalence of opisthorchiasis without the need of stool collection. Using the urine assay, 1 year after the treatment of infected individuals with praziquantel, the prevalence was reduced substantially, but reinfection and new infections reached approximately 6%–10%. Due to the low sensitivity of faecal examination, even by quantitative FECT, data for reinfection and new infection are very limited. Obviously more research is needed for a better and improved diagnosis of opisthorchiasis by molecular, as well as serological techniques. The possibility of the antigen detection in urine has made the detail study transmission dynamics and surveillance much more feasible. Consequently, the results indicate strongly that repeated treatment is required to control and eventually eliminate the parasite in the community. The situation is, to a certain extent, similar to the case of urine antigen detection in schistosomiasis in which schistosomiasis is more prevalent than previously thought creating uncertainty among public health policy makers (Colley et al., 2013). According to the Thai MOPH policy, the cumulative targets for *O. viverrini* screening over the next 10 years are >4 million people. The currently used method for urine assay by ELISA is clearly not practical for handling such a massive screening target. The simplified method, such as using the rapid point of care test format, is required and is being developed for future use in Thailand and for application in other endemic countries.

An important public health aspect of monitoring, control and elimination programs is the detection of *O. viverrini* infections in the snail and fish hosts. Data on snail and fish infections enable the identification of environmental contamination during control and elimination programs.

Fully patent infections in these hosts are detected by parasite recoveries and by molecular parasitological techniques such as PCR or LAMP assays (Parvathi et al., 2008; Duenngai et al., 2008; Hashizume et al., 2017; Arimatsu et al., 2012). Thus, comprehensive diagnosis and detection methods for humans, animal reservoir and intermediate hosts are essential for opsithorchiasis control and elimination programs.

5. DRUG TREATMENT

Chemotherapy, such as praziquantel, has been used for opisthorchiasis treatment and chemotherapeutic control for decades because of its high efficiency, safety margin and reasonable price. The World Health Organization has recommended praziquantel as the drug of choice for opisthorchiasis treatment (WHO, 2006, 2011; WPRO, 2017). However, alternative drugs are being studied. Of these, tribendimidine is the best candidate to date. It shows a promising treatment response (Soukhathammavong et al., 2011). However, an effective treatment alone is not able to control and eliminate opisthorchiasis. The most difficult problem is the traditional and deep-rooted eating behaviour (raw attitudes) and lifestyle of people in the endemic areas (Grundy-Warr et al., 2012). Hence, drug treatment together with integrated control measures is required (Sripa et al., 2015).

5.1 Praziquantel

Praziquantel was discovered in the mid-1970s. From its veterinary use in the beginning, it was developed for trematode and cestode treatment in humans (Lee and Chai, 1985). Praziquantel is a broad-spectrum anthelmintic drug, in tablet form, that has a high safety margin. It is the drug of choice for various food-borne trematodes and schistosomes (WHO, 2006, 2011; WPRO, 2017). The active form of drug increases calcium influx to cells, which leads to the damage of tegument cells and vitelline cells of the parasites. Changing parasitic antigenicity and its metabolism, such as glucose influx, release of lactase, glycogen and decrease of ATP, results from disturbance of intracellular calcium homoeostasis (Day et al., 1992). Subsequently, this biochemical reaction leads to muscle contraction, paralysis and perforation of the tegument causing loss of glucose and amino acid, vacuolization and ultimately rupture of the parasite (Andrew and Harder, 1988; Chan et al., 2012; Doenhoff et al., 2008).

The chemical formula of praziquantel is 2-(cyclohexylcarbonyl)-1,2,3,6,7,11b-hexahydro-4H-pyrazino[2,1-a]isoquinolin-4-one with a molecular weight of 312.3. Although praziquantel has been used for decades, only few pharmacokinetic studies have been reported from humans. The phamacokinetics of praziquantel was summarized by Olliaro et al. (2014). Although it is rapidly and almost completely absorbed, its bioavailability is low and varies between individuals. Oral administration, particularly with food intake, showed a greater pharmacokinetic variability than the intravenous route because of changes in gastric blood flow and pH. The drug can distribute throughout the body and can be concentrated more in the internal organs, particularly the liver and kidneys. Praziquantel can cross the blood–brain barrier and pass into breast milk. It is metabolized in the liver through the CYP system. The main metabolite in humans is trans-4-hydroxypraziquantel. Other metabolites include dehydrogenated and monohydroxylated forms. Excretion of praziquantel mostly occurs through the kidneys.

Praziquantel has been used as 'Preventive Chemotherapy' (WHO, 2006). The recommended dosage used to be 25 mg per kg of body weight, 3 times a day on 2 consecutive days or 40 mg per kg of body weight in a single administration with a repeat interval of 6 or 12 months (WHO, 2011). Currently, the WPRO has just launched a new recommendation of a single dose of 40 mg/kg (WPRO, 2017). The clinical trials of anthelmintic drugs for opisthorchiasis have been carried out mainly in Lao PDR. Sayasone et al. (2017) reported that a single dose of 40 mg/kg in an endemic area is preferable because of its effective outcome and less adverse effects. A flat dose response was observed with praziquantel in light infections and no benefit of multiple doses over a single dose (Sayasone et al., 2017). Although mass drug treatment or universal treatment is included in the WHO guidelines (WHO, 2011), selective chemotherapy or 'Test and Treat' is very effective in Thailand.

Epidemiological and animal studies suggest that repeated infection with *O. viverrini* followed by multiple treatments with praziquantel increases the risk of CCA development (Pinlaor et al., 2004; Kamsa-Ard et al., 2015). As a result, only infected individuals should receive this drug. Therefore, a history of regularly eating raw freshwater fish or fermented fish and/or alcohol consumption will help in determining the target population (WHO, 2011). Praziquantel has different adverse effects, including headache, abdominal pain, dizziness, nausea and urticaria (Keiser and Utzinger, 2004; WHO, 2011). These are usually caused by the substances released from dying worms and systemic effects. These will be more severe in heavy infected individuals and in the population that have not received any previous treatment (WHO, 2011). To prevent adverse effects, postprandial administration after dinner is recommended (personal experience). Drug resistance issues for praziquantel have not been reported in liver fluke; however, there is a report of low cure rate (CR) (29%) for the drug for other parasites such as *C. sinensis* in Vietnam (Tinga et al., 1999).

5.2 Tribendimidine

Tribendimidine is a new candidate for *O. viverrini* treatment. It was first synthesized in China during mid-1980s. Like pyrantel and levamisole, tribendimidine is an L-subtype nicotinic acetylcholine receptor antagonist (Hu et al., 2009; Xiao et al., 2013). It shows a high efficacy against several kinds of helminths (Xiao et al., 2005). In an animal model it decreased *O. viverrini* intensity in hamsters by 95.7% (Keiser et al., 2007). Many clinical studies of tribendimidine conducted in Lao PDR show interesting and promising results. Its efficacy is as good as that of praziquantel. The egg reduction rates (ERRs) of tribendimidine and praziquantel are 99.3% and 98.4%, respectively (Soukhathammavong et al., 2011). A dosage of 100–600 mg/kg showed high therapeutic efficacy in adults and adolescents, while the effective dose for children (8–14 years) was 100 mg/kg. Lower doses give a lower therapeutic efficacy, while the efficacy did not increase when administered with more than 100 mg/kg. Adverse effects could occur 3 h after treatment. These included vertigo, headache, nausea and fatigue. Children reported headache, vertigo and fatigue (Sayasone et al., 2016). Sayasone et al. (2018) conducted a comparative study using an open-label, randomized, noninferiority, phase 2 trial to test the effects of tribendimidine and praziquantel on children (8–14 year) and adults (including adolescents). One oral dose of tribendimidine (200 mg for children, 400 mg for adolescents and adults) and two oral doses of praziquantel (50 mg/kg body weight and 25 mg/kg body weight, 6 h apart) were compared. The results showed that the CR for tribendimidine was slightly lower than that for praziquantel, but that the ERR was not different. The authors prefer tribendimidine because of its efficacy, broad spectrum and safety. However, tribendimidine is a new drug and needs to be approved by a stringent regulatory authority such as FDA.

5.3 Other Treatments

Few studies have been conducted on new trematocidal drugs for opisthorchiasis. These include peroxidic drugs, for example, artemisinins and synthetic trioxolanes. It seems that the therapeutic results were not satisfied. A dose of 400 mg/kg for artesunate and artemether led to worm-burden reductions in hamsters of 77.6% and 65.5%, respectively. Adverse effects were reported when using a dosage over 400 mg/kg (Keiser et al., 2006). In people, the ERR for a mefloquine, artesunate and mefloquine–artesunate combination was below 50%. Adverse effects such as nausea, vomiting, vertigo and anxiety were reported in patients taking mefloquine and mefloquine–artesunate (Soukhathammavong et al., 2011).

Medicinal plants have been studied as possible treatments for opisthorchiasis in vitro and in vivo, such as *Garcinia mangostana* (purple mangosteen) extract (Aukkanimart et al., 2015), *Thunbergia laurifolia* (blue trumpet vine) (Wonkchalee et al., 2013) and curcumin (Charoensuk et al., 2011). It seems that the medicinal plants could be used as a supplement for their antioxidant and antiinflammatory effects, including reduce adverse effects from chemotherapy. A combination or supplement of these with an anthelmintic drug such as praziquantel is now in the research phase (Wonkchalee et al., 2013; Charoensuk et al., 2016).

6. CONCLUSION AND FUTURE RESEARCH

Advances in diagnosis will play a pivotal role in the control and eventual elimination of opisthorchiasis, a risk factor for CCA that affects communities of the lowest socioeconomic status in the Mekong region of Southeast Asia. To achieve the elimination goal for opisthorchiasis and its associated CCA, an effective and practical diagnostic test should be inexpensive and easy to use, preferably in a point of care setting. Of the different methods currently available, parasitology-based techniques are logically challenged and have the major limitation of low diagnostic accuracy, especially for the low-intensity infections in low endemic areas. Recent advances in imaging technology for liver fluke egg identification may help to improve the diagnostic performance of this conventional approach (Jiménez et al., 2016). Difficulties in differentiating between current and past infections and the inability to assess therapeutic responses are major limitations of conventional antibody detection methods, although recent advances such as the development of recombinant-based antibody detection have improved these. Molecular methods yield highly specific results, but more simplification for field application is needed. Among the recent diagnostic developments, the urine antigen assay has shown potential in overcoming some of these challenges, as it can be applied noninvasively. However, further development into a rapid 'point of care' test is necessary such that it can be used for mass screening and so that it is applicable in a resource-poor setting. Moreover, the direct and molecular detection of infected snail and fish intermediate hosts and environmental methods to identify sites of active transmission will provide effective ways of monitoring opisthorchiasis in different geographical locations, especially in Thailand, where disease elimination is now being implemented. Finally, to date, drug treatment for opisthorchiasis has relied solely on praziquantel. Alternative drugs such as tribendimidine have now emerged as potential alternatives, but more tests in endemic areas are required.

Due to the uniqueness of the opisthorchiasis-CCA aetiology in Southeast Asia, which differs markedly from that of Western countries, many new and specific research ventures are required to battle these diseases. A badly needed research outcome is to facilitate the screening of the risk population for CCA. An additional research dimension is to expand and modify the control program being implemented in Thailand into nearby countries where *O. viverrini* is endemic, such that comprehensive programs can be applied and a future opisthorchiasis free community can be achieved.

ACKNOWLEDGEMENTS

This work was supported by Khon Kaen University through CASCAP and the National Research Council of Thailand through the Fluke Free Thailand Project. We wish to thank the Germany Federal Ministry for Education and Research (BMBF, OVISEA project, 01DP15006), the National Research Council of Thailand. We thank Prof Simon Taylor–Robinson, Imperial College, London for his encouragement and enthusiasm.

REFERENCES

Amornpunt, S., Sarasombath, S., Sirisinha, S., 1991. Production and characterization of monoclonal antibodies against the excretory-secretory antigen of the liver fluke (*Opisthorchis viverrini*). Int. J. Parasitol. 21, 421–428.

Andrew, P., Harder, A., 1988. Membrane changes induced by praziquantel. In: Jackson, G.G., Schlumberger, H.D., Zeiler, H.J. (Eds.), Perspectives in Anti Infective Therapy/Current Topics in Infectious Diseases and Clinical Microbiology, vol. 2, pp. 68–75.

Arimatsu, Y., Kaewkes, S., Laha, T., Hong, S.J., Sripa, B., 2012. Rapid detection of *Opisthorchis viverrini* copro-DNA using loop-mediated isothermal amplification (LAMP). Parasitol. Int. 61, 178–182.

Arimatsu, Y., Kaewkes, S., Laha, T., Sripa, B., 2015. Specific diagnosis of *Opisthorchis viverrini* using loop-mediated isothermal amplification (LAMP) targeting parasite microsatellites. Acta Trop. 141, 368–371.

Aukkanimart, R., Boonmars, T., Sriraj, P., Songsri, J., Laummaunwai, P., Waraasawapati, S., Boonyarat, C., Rattanasuwan, P., Boonjaraspinyo, S., 2015. Anthelmintic, anti-inflammatory and antioxidant effect of *Garcinia mangostana* extract in hamster opisthorchiasis. Exp. Parasitol. 154, 5–13.

Aung, W.P.P., Htoon, T.T., Tin, H.H., Thinn, K.K., Sanpool, O., Jongthawin, J., Sadaow, L., Phosuk, I., Rodpai, R., Intapan, P.M., Maleewong, W., 2017. First report and molecular identification of *Opisthorchis viverrini* infection in human communities from Lower Myanmar. PLoS One 12, e0177130.

Boonmekam, D., Namchote, S., Matsuda, H., Kirinoki, K., Sinuon, M., Krailas, D., 2017. Morphological and molecular identification of the liver fluke *Opisthorchis viverrini* in the first intermediate host *Bithynia* snails and its prevalence in Kampong Cham Province, Cambodia. Parasitol. Int. 66, 319–323.

Born-Torrijos, A., Poulin, R., Raga, J.A., Holzer, A.S., 2014. Estimating trematode prevalence in snail hosts using a single-step duplex PCR: how badly does cercarial shedding underestimate infection rates? Parasites Vectors 7, 243.

Buathong, S., Leelayoova, S., Mungthin, M., Naaglor, T., Taamasri, P., Suwannahitatorn, P., Tan-Ariya, P., 2015. Development and evaluation of PCR methods based on cytochrome c oxidase subunit one (*cox*1) and NADH dehydrogenase subunit one gene (*nad*1) to detect *Opisthorchis viverrini* in human fecal samples. Parasitol. Res. 114, 3547–3549.

Buathong, S., Leelayoova, S., Mungthin, M., Ruang-Areerate, T., Naaglor, T., Suwannahitatorn, P., Piyaraj, P., Taamasri, P., Tan-Ariya, P., 2017. Molecular discrimination of *Opisthorchis*-like eggs from residents in a rural community of central Thailand. PLoS Neglected Trop. Dis. 11, e0006030.

Bundhamcharoen, K., Odton, P., Phulkerd, S., Tangcharoensathien, V., 2011. Burden of disease in Thailand: changes in health gap between 1999 and 2004. BMC Publ. Health 11, 53.

Cai, X.Q., Xu, M.J., Wang, Y.H., Qiu, D.Y., Liu, G.X., Lin, A., Tang, J.D., Zhang, R.L., Zhu, X.Q., 2010. Sensitive and rapid detection of *Clonorchis sinensis* infection in fish by loop-mediated isothermal amplification (LAMP). Parasitol. Res. 106, 1379–1383.

Cai, X.Q., Yu, H.Q., Li, R., Yue, Q.Y., Liu, G.H., Bai, J.S., Deng, Y., Qiu, D.Y., Zhu, X.Q., 2014. Rapid detection and differentiation of *Clonorchis sinensis* and *Opisthorchis viverrini* using real-time PCR and high resolution melting analysis. Sci. World J. 2014, 893981.

Caron, Y., Rondelaud, D., Losson, B., 2008. The detection and quantification of a digenean infection in the snail host with special emphasis on *Fasciola* sp. Parasitol. Res. 103, 735–744.

Chai, J.Y., Park, J.H., Han, E.T., Guk, S.M., Shin, E.H., Lin, A., Kim, J.L., Sohn, W.M., Yong, T.S., Eom, K.S., Min, D.Y., Hwang, E.H., Phommmasack, B., Insisiengmay, B., Rim, H.J., 2005. Mixed infections with *Opisthorchis viverrini* and intestinal flukes in residents of Vientiane municipality and Saravane province in Laos. J. Helminthol. 79, 283–289.

Chai, J.Y., Sohn, W.M., Na, B.K., Yong, T.S., Eom, K.S., Yoon, C.H., Hoang, E.H., Jeoung, H.G., Socheat, D., 2014. Zoonotic trematode metacercariae in fish from Phnom Penh and Pursat, Cambodia. Kor. J. Parasitol. 52, 35–40.

Chaicumpa, W., Ybanez, L., Kitikoon, V., Pungpak, S., Ruangkunaporn, Y., Chongsanguan, M., Sornmani, S., 1992. Detection of *Opisthorchis viverrini* antigens in stools using specific monoclonal antibody. Int. J. Parasitol. 22, 527–531.

Chaiyarit, P., Sithithaworn, P., Thuwajit, C., Youngvanit, P., 2011. Detection of salivary antibodies to crude antigens of *Opisthorchis viverrini* in opisthorchiasis and cholangiocarcinoma patients. Clin. Oral Invest. 15, 477–483.

Chan, J.D., Zarowiecki, M., Marchant, J.S., 2012. Ca(2+) channels and Praziquantel: a view from the free world. Parasitol. Int. 12, 161–164.

Chanawong, A., Waikagul, J., Thammapalerd, N., 1990. Detection of shared antigens of human liver flukes *Opisthorchis viverrini* and its snail host, *Bithynia* spp. Trop. Med. Parasitol. 41, 419–421.

Charoensuk, L., Pinlaor, P., Prakobwong, S., Hiraku, Y., Laothong, U., Ruangjirachuporn, W., Yongvanit, P., Pinlaor, S., 2011. Curcumin induces a nuclear factor-erythroid 2-related factor 2-driven response against oxidative and nitrative stress after praziquantel treatment in liver fluke-infected hamsters. Int. J. Parasitol. 41, 615–626.

Charoensuk, L., Pinlaor, P., Wanichwecharungruang, S., Intuyod, K., Vaeteewoottacharn, K., Chaidee, A., Yongvanit, P., Pairojkul, C., Suwannateep, N., Pinlaor, S., 2016. Nanoencapsulated curcumin and praziquantel treatment reduces periductal fibrosis and attenuates bile canalicular abnormalities in *Opisthorchis viverrini*-infected hamsters. Nanomedicine 12, 21–32.

Chen, D., Chen, J., Huang, J., Chen, X., Feng, D., Liang, B., Che, Y., Liu, X., Zhu, C., Li, X., Shen, H., 2010. Epidemiological investigation of *Clonorchis sinensis* infection in freshwater fishes in the Pearl River Delta. Parasitol. Res. 107, 835–839.

Colley, D.G., Binder, S., Campbell, C., King, C.H., Tchuem Tchuenté, L.A., N'Goran, E.K., Erko, B., Karanja, D.M., Kabatereine, N.B., van Lieshout, L., Rathbun, S., 2013. A five-country evaluation of a point-of care circulating cathodic antigen urine assay for the prevalence of Schistosoma mansoni. Am. J. Trop. Med. Hyg. 88, 426–432.

Dao, T.H., Nguyen, T.G., Victor, B., Gabriël, S., Dorny, P., 2014. *Opisthorchis viverrini*-like liver fluke in birds from Vietnam: morphological variability and rDNA/mtDNA sequence confirmation. J. Helminthol. 88, 441–446.

Dao, T.H., Abatih, E.N., Nguyen, T.T., Tran, T.T., Gabriël, S., Smit, S., Le, P.N., Dorny, P., 2016. Prevalence of *Opisthorchis viverrini*-like fluke infection in ducks in Binh Dinh province, central Vietnam. Kor. J. Parasitol. 54, 357–361.

Dao, H.T.T., Dermauw, V., Gabriël, S., Suwannatrai, A., Tesana, S., Nguyen, G.T.T., Dorny, P., 2017. *Opisthorchis viverrini* infection in the snail and fish intermediate hosts in central Vietnam. Acta Trop. 170, 120–125.

Day, T.A., Bennett, J.L., Pax, R.A., 1992. Praziquantel the enigmatic antiparasitic. Parasitol. Today 8, 342–344.

De, N.V., Le, T.H., 2011. Human infections of fish-borne trematodes in vietnam: prevalence and molecular specific identification at an endemic commune in Nam Dinh province. Exp. Parasitol. 129, 355–361.

Doenhoff, M.J., Cioli, D., Utzinger, J., 2008. Praziquantel: mechanisms of action, resistance and new derivatives for schistosomiasis. Curr. Opin. Infect. Dis. 21, 659–667.

Duenngai, K., Sithithaworn, P., Rudrappa, U.K., Iddya, K., Laha, T., Stensvold, C.R., Strandgaard, H., Johansen, M.V., 2008. Improvement of PCR for detection of *Opisthorchis viverrini* DNA in human stool samples. J. Clin. Microbiol. 46, 366–368.

Duenngai, K., Boonmars, T., Sithithaworn, J., Sithithaworn, P., 2013. Diagnosis of early infection and post chemotherapeutic treatment by copro-DNA detection in experimental opisthorchiasis. Parasitol. Res. 112, 271–278.

Elkins, D.B., Haswell-Elkins, M.R., Mairiang, E., Mairiang, P., Sithithaworn, P., Kaewkes, S., Bhudhisawasdi, V., Uttaravichien, T., 1990. A high frequency of hepatobiliary disease and suspected cholangiocarcinoma associated with heavy *Opisthorchis viverrini* infection in a small community in north-east Thailand. Trans. R. Soc. Trop. Med. Hyg. 84, 715–719.

Elkins, D.B., Sithithaworn, P., Haswell-Elkins, M., Kaewkes, S., Awacharagan, P., Wongratanacheewin, S., 1991. *Opisthorchis viverrini*: relationships between egg counts, worms recovered and antibody levels within an endemic community in northeast Thailand. Parasitology 102, 283–288.

Enes, J.E., Wages, A.J., Malone, J.B., Tesana, S., 2010. Prevalence of *Opisthorchis viverrini* infection in the canine and feline hosts in three villages, Khon Kaen Province, northeastern Thailand. Southeast Asian J. Trop. Med. Publ. Health 41, 36–42.

Ficetola, G.F., Miaud, C., Pompanon, F., Taberlet, P., 2008. Species detection using environmental DNA from water samples. Biol. Lett. 4, 423–425.

Grundy-Warr, C., Andrews, R.H., Sithithaworn, P., Petney, T.N., Sripa, B., Laithavewat, L., Ziegler, A.D., 2012. Raw attitudes, wetland cultures, life-cycles: socio-cultural dynamics relating to *Opisthorchis viverrini* in the Mekong Basin. Parasitol. Int. 61, 65–70.

Hashizume, H., Sato, M., Sato, M.O., Ikeda, S., Yoonuan, T., Sanguankiat, S., Pongvongsa, T., Moji, K., Minamoto, T., 2017. Application of environmental DNA analysis for the detection of *Opisthorchis viverrini* DNA in water samples. Acta Trop. 169, 1–7.

Haswell-Elkins, M.R., Sithithaworn, P., Mairiang, E., Elkins, D.B., Wongratanacheewin, S., Kaewkes, S., Mairiang, P., 1991. Immune responsiveness and parasite-specific antibody levels in human hepatobiliary disease associated with *Opisthorchis viverrini* infection. Clin. Exp. Immunol. 84, 213–218.

Haswell-Elkins, M.R., Satarug, S., Elkins, D.B., 1992. *Opisthorchis viverrini* infection in northeast Thailand and its relationship to cholangiocarcinoma. J. Gastroenterol. Hepatol. 7, 538–548.

Hong, S.J., Shin, J.K., Kang, S.Y., Ryu, J.R., 2003. Ultrastructural localization of phosphoglycerate kinase in adult *Clonorchis sinensis*. Parasitol. Res. 90, 369–371.

Hu, Y., Xiao, S.H., Aroian, R.V., 2009. The new anthelmintic tribendimidine is an L-type (levamisole and pyrantel) nicotinic acetylcholine receptor agonist. PLoS Neglected Trop. Dis. 3, e499.

IARC, 2012. Biological agents: a review of human carcinogens. IARC (Int. Agency Res. Cancer) Monogr. Eval. Carcinog. Risks Hum. 100B, 341–370.

Intapan, P.M., Thanchomnang, T., Lulitanond, V., Phongsaskulchoti, P., Maleewong, W., 2008. Real-time fluorescence resonance energy transfer PCR with melting curve analysis for the detection of *Opisthorchis viverrini* in fish intermediate hosts. Vet. Parasitol. 157, 65–71.

Intapan, P.M., Thanchomnang, T., Lulitanond, V., Pongsaskulchoti, P., Maleewong, W., 2009. Rapid molecular detection of *Opisthorchis viverrini* in human fecal samples by real-time polymerase chain reaction. Am. J. Trop. Med. Hyg. 81, 917–920.

Janwan, P., Intapan, P.M., Thanchomnang, T., Lulitanond, V., Anamnart, W., Maleewong, W., 2011. Rapid detection of *Opisthorchis viverrini* and *Strongyloides stercoralis* in human fecal samples using a duplex real-time PCR and melting curve analysis. Parasitol. Res. 109, 1593–1601.

Jawalagatti, V., Dadas, S.A., Mishra, S., Gudewar, J., Sawale, G.K., Gatne, M.L., 2016. A rare abattoir report on heavy infection of pigs with *Opisthorchis tenuicollis*. Isr. J. Vet. Med. 71, 32–36.

Jiménez, B., Maya, C., Velásquez, G., Torner, F., Arambula, F., Barrios, J.A., Velasco, M., 2016. Identificaiton and quantification of pathogenic helminth eggs using a digital image system. Exp. Parasitol. 166, 164–172.

Johansen, M.V., Lier, T., Sithithaworn, P., 2015. Towards improved diagnosis of neglected zoonotic trematodes using a One Health approach. Acta Trop. 141, 161–169.

Kaewkes, S., Kaewkes, W., Boonmars, T., Sripa, B., 2012. Effect of light intensity on *Opisthorchis viverrini* cercarial shedding levels from Bithynia snails—a preliminary study. Parasitol. Int. 61, 46–48.

Kaewkong, W., Intapan, P.M., Sanpool, O., Janwan, P., Thanchomnang, T., Laummaunwai, P., Lulitanond, V., Doanh, P.N., Maleewong, W., 2013. Molecular differentiation of *Opisthorchis viverrini* and *Clonorchis sinensis* eggs by multiplex real-time PCR with high resolution melting analysis. Kor. J. Parasitol. 51, 689–694.

Kaewpitoon, N., Laha, T., Kaewkes, S., Yongwanit, P., Brindley, P.J., Loukas, A., Sripa, B., 2008. Characterization of cysteine protease from the carcinogenic liver fluke, *Opisthorchis viverrini*. Parasitol. Res. 102, 757–764.

Kamsa-Ard, S., Luvira, V., Pugkhem, A., Luvira, V., Thinkhamrop, B., Suwanrungruang, K., Bhudhisawasdi, V., 2015. Association between praziquantel treatment and cholangiocarcinoma: a hospital-based matched case-control study. BMC Canc. 15, 776.

Kang, S., Sultana, T., Loktev, V.B., Wongratanacheewin, S., Sohn, W.M., Eom, K.S., Park, J.K., 2008. Molecular identification and phylogenetic analysis of nuclear rDNA sequences among three opisthorchid liver fluke species (Opisthorchiidae: Trematoda). Parasitol. Int. 57, 191–197.

Keiser, J., Utzinger, J., 2004. Chemotherapy for major food-borne trematodes: a review. Expert Opin. Pharmacother. 5, 1711–1726.

Keiser, J., Shu-Hua, X., Chollet, J., Tanner, M., Utzinger, J., 2007. Evaluation of the in vivo activity of tribendimidine against *Schistosoma mansoni, Fasciola hepatica, Clonorchis sinensis,* and Opisthorchis viverrini. Antimicrob. Agents Chemother. 51, 1096–1098.

Keiser, J., Shu-Hua, X., Jian, X., Zhen-San, C., Odermatt, P., Tesana, S., Tanner, M., Utzinger, J., 2006. Effect of artesunate and artemether against *Clonorchis sinensis* and *Opisthorchis viverrini* in rodent models. Int. J. Antimicrob. Agents 28, 370–373.

Khuntikeo, N., Loilome, W., Thinkhamrop, B., Chamadol, N., Yongvanit, P., 2016. A comprehensive public health conceptual framework and strategy to effectively combat cholangiocarcinoma in Thailand. PLoS Neglected Trop. Dis. 10, e0004293.

Kiatsopit, N., Sithithaworn, P., Saijuntha, W., Boonmars, T., Tesana, S., Sithithaworn, J., Petney, T.N., Andrews, R.H., 2012. Exceptionally high prevalence of infection of *Bithynia siamensis goniomphalos* with *Opisthorchis viverrini* cercariae in different wetlands in Thailand and Lao PDR. Am. J. Trop. Med. Hyg. 86, 464–469.

Kiatsopit, N., Sithithaworn, P., Saijuntha, W., Pitaksakulrat, O., Petney, T.N., Webster, J.P., Andrews, R.H., 2014. Analysis of the population genetics of *Opisthorchis viverrini* sensu lato in the Nam Ngum River wetland, Lao PDR, by multilocus enzyme electrophoresis. Parasitol. Res. 113, 2973–2981.

Kottelat, M., 2013. The fishes of the inland waters of Southeast Asia: a catalogue and core bibliography of the fishes known to occur in freshwaters, mangroves and estuaries. Raffles Bull. Zool. 27, 1–663.

Laha, T., Sripa, J., Sripa, B., Pearson, M., Tribolet, L., Kaewkes, S., Sithithaworn, P., Brindley, P.J., Loukas, A., 2008. Asparaginyl endopeptidase from the carcinogenic liver fluke, *Opisthorchis viverrini*, and its potential for serodiagnosis. Int. J. Infect. Dis. 12, e49–e59.

Lamaningao, P., Kanda, S., Laimanivong, S., Shimono, T., Darcy, A.W., Phyaluanglath, A., Mishima, N., Nishiyama, T., 2017. Development of a PCR assay for diagnosing trematode (*Opisthorchis* and *Haplorchis*) infections in human stools. Am. J. Trop. Med. Hyg. 96, 221–228.

Lan-Anh, N.T., Phuong, N.T., Murrell, K.D., Johansen, M.V., Dalsgaard, A., Thu, L.T., Kim-Chi, T.T., Thamsborg, S.M., 2009. Animal reservoir hosts and fish-borne zoonotic trematode infections on fish farms, Vietnam. Emerg. Infect. Dis. 15, 540–546.

Laoprom, N., Sithithaworn, P., Andrews, R.H., Ando, K., Laha, T., Klinbunga, S., Webster, J.P., Petney, T.N., 2012. Population genetic structuring in *Opisthorchis viverrini* over various spatial scales in Thailand and Lao PDR. PLoS Neglected Trop. Dis. 6, e1906.
Laoprom, N., Laithavewat, L., Kopolrat, K., Kiatsopit, N., Kaewkes, S., Chantaluk, S., Mongkolsin, C., Chanmaha, B., Andrews, R.H., Petney, T.N., Sithithaworn, P., 2016a. Evaluation of a commercial stool concentrator kit compared to direct smear and formalin-ethyl acetate concentration methods for diagnosis of parasite infection with special reference to *Opisthorchis viverrini* sensu lato in Thailand. Southeast Asian J. Trop. Med. Publ. Health 47, 890–900.
Laoprom, N., Kiatsopit, N., Sithithaworn, P., Kopolrat, K., Andrews, R.H., Petney, T.N., 2016b. Cercarial emergence patterns for *Opisthorchis viverrini* sensu lato infecting *Bithynia siamensis goniomphalos* from Sakon Nakhon Province, Thailand. Parasitol. Res. 115, 3313–3321.
Lee, S.H., Chai, J.Y., 1985. Praziquantel in the treatment of trematode and cestode infections. J. Korean Soc. Chemother. 3, 95–118.
Le, T.H., Nguyen, N.T., Truong, N.H., De, N.V., 2012. Development of mitochondrial loop-mediated isothermal amplification for detection of the small liver fluke *Opisthorchis viverrini* (Opisthorchiidae; Trematoda; Platyhelminthes). J. Clin. Microbiol. 50, 1178–1184.
Le, T.H., Van De, N., Blair, D., Sithithaworn, P., McManus, D.P., 2006. *Clonorchis sinensis* and *Opisthorchis viverrini*: development of a mitochondrial-based multiplex PCR for their identification and discrimination. Exp. Parasitol. 112, 109–114.
Li, S., Shin, J.G., Cho, P.Y., Kim, T.I., Hong, S.T., Hong, S.J., 2011. Multiple recombinant antigens of *Clonorchis sinensis* for serodiagnosis of human clonorchiasis. Parasitol. Res. 108, 1295–1302.
Llewellyn, S., Inpankaew, T., Nery, S.V., Gray, D.J., Verweij, J.J., Clements, A.C., Gomes, S.J., Traub, R., McCarthy, J.S., 2016. Application of a multiplex quantitative PCR to assess prevalence and intensity of intestinal parasite infections in a controlled clinical trial. PLoS Neglected Trop. Dis. 10, e0004380.
Lovis, L., Mak, T.K., Phongluxa, K., Soukhathammavong, P., Sayasone, S., Akkhavong, K., Odermatt, P., Keiser, J., Felger, I., 2009. PCR diagnosis of *Opisthorchis viverrini* and *Haplorchis taichui* infections in a Lao community in an area of endemicity and comparison of diagnostic methods for parasitological field surveys. J. Clin. Microbiol. 47, 1517–1523.
Mairiang, E., Elkins, D.B., Mairiang, P., Chaiyakum, J., Chamadol, N., Laopaiboon, V., Posri, S., Sithithaworn, P., Haswell-Elkins, M., 1992. Relationship between intensity of *Opisthorchis viverrini* infection and hepatobiliary disease detected by ultrasonography. J. Gastroenterol. Hepatol. 7, 17–21.
Maleewong, W., Intapan, P.M., Wongkham, C., Wongsaroj, T., Kowsuwan, T., Pumdonming, W., Pongsakulchoti, P., Kitikoon, V., 2003. Detection of *Opisthorchis viverrini* in experimentally infected bithynid snails and cyprinoid fishes by a PCR-based method. Parasitology 126, 63–67.
Meurs, L., Polderman, A.M., Vinkeles Melchers, N.V.S., Brienen, E.A., Verweij, J.J., Groosjohan, B., Mendes, F., Mechendura, M., Hepp, D.H., Langenberg, M.C., Edelenbosch, R., Polman, K., van Lieshout, L., 2017. Diagnosing polyparasitisim in a high-prevalence setting in Beira, Mozambique: detection of intestinal parasites in fecal samples by microscopy and real-time PCR. PLoS Neglected Trop. Dis. 11, e0005310.
Minamoto, K., Mascie-Taylor, C.G., Karim, E., Moji, K., Rahman, M., 2012. Short- and long-term impact of health education in improving water supply, sanitation and knowledge about intestinal helminths in rural Bangladesh. Publ. Health 126, 437–440.

Ngern-klun, R., Sukontason, K.L., Tesana, S., Sripakdee, D., Irvine, K.N., Sukontason, K., 2006. Field investigation of Bithynia funiculata, intermediate host of *Opisthorchis viverrini* in northern Thailand. Southeast Asian J. Trop. Med. Publ. Health 37, 662–672.

Olliaro, P., Delgado-Romero, P., Keiser, J., 2014. The little we know about the pharmacokinetics and pharmacodynamics of praziquantel (racemate and R-enantiomer). J. Antimicrob. Chemother. 69, 863–870.

Pakdee, W., Waikagul, J., Kalambaheti, T., Ito, A., Dekumyoy, P., 2010. Iso-electricfocusing of *Bithynia* snail antigens for IgG- and IgG(1-4)-ELISA detection of human opisthorchiasis. Southeast Asian J. Trop. Med. Publ. Health 41, 813–820.

Pande, B.P., Shukla, R.P., 1974. Experimentally induced *Opisthorchis caninus* infection in hamsters & rhesus monkey with a note on the role of larval digeneans in fisheries production. Indian J. Exp. Biol. 12, 184–191.

Parvathi, A., Sanath Kumar, H., Kenchanna Prakasha, B., Lu, J., Xu, X., Hu, W., Feng, Z., Karunasagar, I., Karunasagar, I., 2007. *Clonorchis sinensis*: development and evaluation of a nested polymerase chain reaction (PCR) assay. Exp. Parasitol. 115, 291–295.

Parvathi, A., Umesha, K.R., Kumar, S., Sithithaworn, P., Karunasagar, I., Karunasagar, I., 2008. Development and evaluation of a polymerase chain reaction (PCR) assay for the detection of *Opisthorchis viverrini* in fish. Acta Trop. 107, 13–16.

Pasuralertsakul, S., Ngrenngarmlert, W., Sripochang, S., Khantiyanan, N., Akkarangamsiri, O., 2005. Methylene blue staining method for identification of *Opisthorchis viverini* egg. Southeast Asian J. Trop. Med. Publ. Health 36, 107–109.

Petney, T., Sithithaworn, P., Andrews, R., Kiatsopit, N., Tesana, S., Grundy-Warr, C., Ziegler, A., 2012. The ecology of the *Bithynia* first intermediate hosts of *Opisthorchis viverrini*. Parasitol. Int. 61, 38–45.

Phan, V.T., Ersboll, A.K., Nguyen, K.V., Madsen, H., Dalsgaard, A., 2010a. Farm-level risk factors for fish-borne zoonotic trematode infection in integrated small-scale fish farms in northern Vietnam. PLoS Neglected Trop. Dis. 4, e742.

Phan, V.T., Ersbøll, A.K., Nguyen, T.T., Nguyen, K.V., Nguyen, H.T., Murrell, D., Dalsgaard, A., 2010b. Freshwater aquaculture nurseries and infection of fish with zoonotic trematodes, Vietnam. Emerg. Infect. Dis. 16, 1905–1909.

Phongsasakulchoti, P., Sri-aroon, P., Kerdpuech, Y., 2005. Emergence of *Opisthorchis viverrini* cercariae from naturally infected *Bithynia (Digoniostoma) siamensis goniomphalos*. Southeast Asian J. Trop. Med. Publ. Health 36, 189–191.

Pinlaor, S., Ma, N., Hiraku, Y., Yongvanit, P., Semba, R., Oikawa, S., Murata, M., Sripa, B., Sithithaworn, P., Kawanishi, S., 2004. Repeated infection with *Opisthorchis viverrini* induces accumulation of 8-nitroguanine and 8-oxo-7,8-dihydro-2'-deoxyguanine in the bile duct of hamsters via inducible nitric oxide synthase. Carcinogenesis 25, 1535–1542.

Pinlaor, P., Kaewpitoon, N., Laha, T., Sripa, B., Kaewkes, S., Morales, M.E., Mann, V.H., Parriott, S.K., Suttiprapa, S., Robinson, M.W., To, J., Dalton, J.P., Loukas, A., Brindley, P.J., 2009. Cathepsin F cysteine protease of the human liver fluke, *Opisthorchis viverrini*. PLoS Neglected Trop. Dis. 3, e398.

Pinlaor, S., Onsurathum, S., Boonmars, T., Pinlaor, P., Hongsrichan, N., Chaidee, A., Haonon, O., Limviroj, W., Tesana, S., Kaewkes, S., Sithithaworn, P., 2013. Distribution and abundance of *Opisthorchis viverrini* metacercariae in cyprinid fish in Northeastern Thailand. Kor. J. Parasitol. 51, 703–710.

Pitaksakulrat, O., Sithithaworn, P., Laoprom, N., Laha, T., Petney, T.N., Andrews, R.H., 2013. A cross-sectional study on the potential transmission of the carcinogenic liver fluke *Opisthorchis viverrini* and other fishborne zoonotic trematodes by aquaculture fish. Foodborne Pathogens Disease 10, 35–41.

Poopyruchpong, N., Viyanant, V., Upatham, E.S., Srivatanakul, P., 1990. Diagnosis of opisthorchis by enzyme-linked immunosorbent assay using partially purified antigens. Asian Pac. J. Allergy Immunol. 8, 27–31.

Pumidonming, W., Salman, D., Gronsang, D., Abdelbaset, A.E., Sangkaeo, K., Kawazu, S.I., Igarashi, M., 2016. Prevalence of gastrointestinal parasites of zoonotic significance in dogs and cats in lower Northern Thailand. J. Vet. Med. Sci. 78, 1779–1784.

Radomyos, P., Radomyos, B., Tungtrongchitr, A., 1994. Multi-infection with helminths in adults from northeast Thailand as determined by post-treatment fecal examination of adult worms. Trop. Med. Parasitol. 45, 133–135.

Rahman, S.M., Bae, Y.M., Hong, S.T., Choi, M.H., 2011. Early detection and estimation of infection burden by real-time PCR in rats experimentally infected with *Clonorchis sinensis*. Parasitol. Res. 109, 297–303.

Ramsay, R.J., Sithithaworn, P., Prociv, P., Moorhouse, D.E., Methaphat, C., 1989. Density-dependent fecundity of *Opisthorchis viverrini* in humans, based on faecal recovery of flukes. Trans. R. Soc. Trop. Med. Hyg. 83, 241–242.

Rattanachan, S., Grams, R., Tesana, S., Smooker, P.M., Grams, S.V., 2014. *Opisthorchis viverrini*: analysis of the sperm-specific rhophilin associated tail protein 1-like. Acta Trop. 140, 34–40.

Riganti, M., Pungpak, S., Harinasuta, T., Bunnag, D., 1988. Transient rise in *Opisthorchis* egg count after praziquantel treatment. Southeast Asian J. Trop. Med. Publ. Health 19, 335–336.

Riganti, M., Pungpak, S., Punpoowong, B., Bunnag, D., Harinasuta, T., 1989. Human pathology of Opisthorchis viverrini infection: a comparison of adults and children. Southeast Asian J. Trop. Med. Publ. Health 20, 95–100.

Ruangsittichai, J., Viyanant, V., Vichasri-Grams, S., Sobhon, P., Tesana, S., Upatham, E.S., Hofmann, A., Korge, G., Grams, R., 2006. *Opisthorchis viverrini*: identification of a glycine-tyrosine rich eggshell protein and its potential as a diagnostic tool for human opisthorchiasis. Int. J. Parasitol. 36, 1329–1339.

Saichua, P., Sithithaworn, P., Jariwala, A.R., Diemert, D.J., Sithithaworn, J., Sripa, B., Laha, T., Mairiang, E., Pairojkul, C., Periago, M.V., Khuntikeo, N., Mulvenna, J., Bethony, J.M., 2013. Microproteinuria during *Opisthorchis viverrini* infection: a biomarker for advanced renal and hepatobiliary pathologies from chronic opisthorchiasis. PLoS Neglected Trop. Dis. 7, e2228.

Saijuntha, W., Sithithaworn, P., Wongkham, S., Laha, T., Pipitgool, V., Tesana, S., Chilton, N.B., Petney, T.N., Andrews, R.H., 2007. Evidence of a species complex within the food-borne trematode *Opisthorchis viverrini* and possible co-evolution with their first intermediate hosts. Int. J. Parasitol. 37, 695–703.

Saijuntha, W., Sithithaworn, P., Kaitsopit, N., Andrews, R.H., Petney, T.N., 2014. Liver flukes: *Clonorchis* and *Opisthorchis*. Adv. Exp. Med. Biol. 766, 153–199.

Sanpool, O., Intapan, P.M., Thanchomnang, T., Janwan, P., Lulitanond, V., Doanh, P.N., Van Hien, H., Dung do, T., Maleewong, W., Nawa, Y., 2012. Rapid detection and differentiation of *Clonorchis sinensis* and *Opisthorchis viverrini* eggs in human fecal samples using a duplex real-time fluorescence resonance energy transfer PCR and melting curve analysis. Parasitol. Res. 111, 89–96.

Sato, M., Thaenkham, U., Dekumyoy, P., Waikagul, J., 2009. Discrimination of *O. viverrini, C. sinensis, H. pumilio* and *H. taichui* using nuclear DNA-based PCR targeting ribosomal DNA ITS regions. Acta Trop. 109, 81–83.

Sato, M., Pongvongsa, T., Sanguankiat, S., Yoouan, T., Kobayashi, J., Boupha, B., Nishimoto, F., Moji, K., Sato, M.O., Waikagul, J., 2015. Pattern of trematode infections of *Opisthorchis viverrini* (Opisthorchiidae) and *Haplorchis taichui* (Heterophyidae) in human populations from two villages in Savannakhet Province, Lao PDR. J. Helminthol. 89, 439–445.

Sawangsoda, P., Sithithaworn, J., Tesana, S., Pinlaor, S., Boonmars, T., Mairiang, E., Yongvanit, P., Duenngai, K., Sithithaworn, P., 2012. Diagnostic values of parsite-specific antibody detections in saliva and urine in comparison with serum in opisthorchiasis. Parasitol. Int. 61, 196–202.

Sayasone, S., Utzinger, J., Akkhavong, K., Odermatt, P., 2015. Multiparasitism and intensity of helminth infections in relation to symptoms and nutritional status among children: a cress-sectional study in southern Lao People's Democratic Republic. Acta Trop. 141, 322–331.
Sayasone, S., Odermatt, P., Vonghachack, Y., Xayavong, S., Senggnam, K., Duthaler, U., Akkhavong, K., Hattendorf, J., Keiser, J., 2016. Efficacy and safety of tribendimidine against *Opisthorchis viverrini*: two randomized, parallel-group, single-blind, dose-ranging, phase 2 trials. Lancet Infect. Dis. 16, 1145–1153.
Sayasone, S., Meister, I., Andrews, J.R., Odermatt, P., Vonghachack, Y., Xayavong, S., Senggnam, K., Phongluxa, K., Hatterdorf, J., Bogoch, I.I., Keiser, J., 2017. Efficacy and safety of paraziquantel against light infections of *Opisthorchis viverrini*: a randomized parallel single-blind dose-ranging trial. Clin. Infect. Dis. 64, 451–458.
Sayasone, S., Odermatt, P., Vonghachack, Y., Xayavong, S., Senggnam, K., Duthaler, U., Akkhavong, K., Hattendorf, J., Keiser, J., 2018. Efficacy and safety of tribendimidine versus praziquantel against *Opisthorchis viverrini* in Laos: an open-lable, randomised, non-inferiority, phase 2 trial. Lancet Infect. Dis. 18, 155–161.
Silakit, R., Loilome, W., Yongvanit, P., Thongchot, S., Sithithaworn, P., Boonmars, T., Koonmee, S., Titapun, A., Khuntikeo, N., Chamadol, N., Techasen, A., Namwat, N., 2017. Urinary microRNA-192 and microRNA-21 as potential indicators for liver fluke-associated cholangiocarcinoma risk group. Parasitol. Int. 66, 479–485.
Sirisinha, S., Sahassananda, D., Bunnag, D., Rim, H.J., 1990. Immunological analysis of *Opisthorchis* and *Clonorchis* antigens. J. Helminthol. 64, 133–138.
Sirisinha, S., Chawengkirttikul, R., Sermswan, R., Amornpant, S., Mongkolsuk, S., Panyim, S., 1991. Detection of *Opisthorchis viverrini* by monoclonal antibody-based ELISA and DNA hybridization. Am. J. Trop. Med. Hyg. 44, 140–145.
Sirisinha, S., Chawengkirttikul, R., Haswell-Elkins, M.R., Elkins, D.B., Kaewkes, S., Sithithaworn, P., 1995. Evaluation of a monoclonal antibody-based enzyme linked immunosorbent assay for the diagnosis of *Opisthorchis viverrini* infection in an endemic area. Am. J. Trop. Med. Hyg. 52, 521–524.
Sithithaworn, P., Haswell-Elkins, M., 2003. Epidemiology of *Opisthorchis viverrini*. Acta Trop. 88, 187–194.
Sithithaworn, P., Pipitgool, V., Srisawangwong, T., Elkins, D.B., Haswell-Elkins, M.R., 1997. Seasonal variation of *Opisthorchis viverrini* infection in cyprinoid fish in north-east Thailand: implications for parasite control and food safety. Bull. World Health Organ. 75, 125–131.
Sithithaworn, P., Tesana, S., Pipitgool, V., Kaewkes, S., Pairojkul, C., Sripa, B., Paupairoj, A., Thaiklar, A., 1991. Relationship between faecal egg count and worm burden of *Opisthorchis viverrini* in human autopsy cases. Parasitology 102, 277–281.
Sithithaworn, P., Andrews, R.H., Petney, T.N., Saijuntha, W., Laoprom, N., 2012. The systematics and population genetics of *Opisthorchis viverrini* sensu lato: implications in parasite epidemiology and bile duct cancer. Parasitol. Int. 61, 32–37.
Sithithaworn, P., Yongvanit, P., Duenngai, K., Kiatsopit, N., Pairojkul, C., 2014. Roles of liver fluke infection as risk factor for cholangiocarcinoma. J. Hepato-Biliary-Pancreatic Sci. 21, 301–308.
Soukhathammavong, P., Odermatt, P., Sayasone, S., Vonghachack, Y., Vounatsou, P., Hatz, C., Akkhavong, K., Keiser, J., 2011. Efficacy and safety of mefloquine, artesunate, mefloquine-artesunate, tribendimidine, and praziquantel in patients with *Opisthorchis viverrini*: a randomised, exploratory, open-label, phase 2 trial. Lancet Infect. Dis. 11, 110–118.
Sri-Aroon, P., Intapan, P.M., Lohachit, C., Phongsasakulchoti, P., Thanchomnang, T., Lulitanond, V., Hiscox, A., Phompida, S., Sananikhom, P., Maleewong, W., Brey, P.T., 2011. Molecular evidence of Opisthorchis viverrini in infected bithyniid snails in the Lao People's Democratic Republic by specific hybridization probe-based real-time fluorescence resonance energy transfer PCR method. Parasitol. Res. 108, 973–978.

Sripa, B., Brindley, P.J., Mulvenna, J., Laha, T., Mairiang, E., Bethony, J.M., Loukas, A., 2012a. The tumorigenic liver fluke *Opisthorchis viverrini*—multiple pathways to cancer. Trends Parasitol. 28, 395–407.

Sripa, J., Brindley, P.J., Sripa, B., Loukas, A., Kaewkes, S., Laha, T., 2012b. Evaluation of liver fluke recombinant cathepsin B-1 protease as a serodiagnostic antigen for human opisthorchiasis. Parasitol. Int. 61, 191–195.

Sripa, B., Tangkawattana, S., Laha, T., Kaewkes, S., Mallory, F.F., Smith, J.F., Wilcox, B.A., 2015. Toward integrated opisthorchiasis control in northeast Thailand: the Lawa project. Acta Trop. 141, 361–367.

Srisawangwong, T., Sithithaworn, P., Tesana, S., 1997. Metacercariae isolated from cyprinoid fishes in Khon Kaen District by digestion technic. Southeast Asain J. Trop. Med. Public Health 28, 224–226.

Stensvold, C.R., Saijuntha, W., Sithithaworn, P., Wongratanacheewin, S., Strandgaard, H., Ornbjerg, N., Johansen, M.V., 2006. Evaluation of PCR based coprodiagnosis of human opisthorchiasis. Acta Trop. 97, 26–30.

Studer, A., Poulin, R., 2012. Seasonal dynamics in an intertidal mudflat: the case of a complex trematode life cycle. Mar. Ecol. Prog. Ser. 455, 79–93.

Stunkard, H.W., Hinchliffe, M.C., 1952. The morphology and life-history of *Microbilhazia variglandis* (Miller and Northup, 1962) Stunkard and Hinchliffe, 1951, avian blood-flukes whose larvae cause "swimmer's itch" of ocean beaches. J. Parasitol. 38, 248–265.

Sukontason, K., Piangjai, S., Muangyimpong, Y., Sukontason, K., Methanitikorn, R., Chaithong, U., 1999. Prevalence of trematode metacercariae in cyprinoid fish of Ban Pao district, Chiang Mai Province, northern Thailand. Southeast Asian J. Trop. Med. Publ. Health 30, 365–370.

Suksumek, N., Leelawat, K., Leelawat, S., Russell, B., Lek-Uthai, U., 2008. TaqMan real-time PCR assay for specific detection of *Opisthorchis viverrini* DNA in Thai patients with hepatocellular carcinoma and cholangiocarcinoma. Exp. Parasitol. 119, 217–224.

Sun, J., Xu, J., Liang, P., Mao, Q., Huang, Y., Lv, X., Deng, C., Liang, C., de Hoog, G.S., Yu, X., 2011. Molecular identification of *Clonorchis sinensis* and discrimination with other opisthorchid liver fluke species using multiple Ligation-depended Probe Amplification (MLPA). Parasites Vectors 4, 98.

Tantrawatpan, C., Intapan, P.M., Thanchomnang, T., Sanpool, O., Janwan, P., Lulitanond, V., Sadaow, L., Maleewong, W., 2014. Development of a PCR assay and pyrosequencing for identification of important human fish-borne trematodes and its potential use for detection in fecal specimens. Parasites Vectors 7, 8.

Teimoori, S., Arimatsu, Y., Laha, T., Kaewkes, S., Serrerak, P., Tangkawattana, S., Brindley, P.J., Sripa, B., 2015. Immunodiagnosis of opisthorchiasis using parasite cathepsin F. Parasitol. Res. 114, 4571–4578.

Teimoori, S., Arimatsu, Y., Laha, T., Kaewkes, S., Sereerak, P., Sripa, M., Tangkawattana, S., Brindley, P.J., Sripa, B., 2017. Chicken IgY-based coproantigen capture ELISA for diagnosis of human opisthorchiasis. Parasitol. Int. 66, 443–447.

Tesana, S., Srisawangwong, T., Sithithaworn, P., Itoh, M., Phumchaiyothin, R., 2007. The ELISA-based detection of anti-*Opisthorchis viverrini* IgG and IgG4 in samples of human urine and serum from an endemic area of north-eastern Thailand. Ann. Trop. Med. Parasitol. 101, 585–591.

Thaenkham, U., Nuantanong, S., Vonghachack, Y., Yoonuan, T., Sanguankiat, S., Dekumyoy, P., Prommasack, B., Kobayashi, J., Waikagul, J., 2011. Discovery of *Opisthorchis lobatus* (Trematoda: Opisthorchiidae): a new record of small liver flukes in the greater Mekong sub-region. J. Parasitol. 97, 1152–1158.

Thi Phung, L., Loukas, A., Brindley, P.J., Sripa, B., Laha, T., 2014. Retrotransposon OV-RTE-1 from the carcinogenic liver flukes *Opisthorchis viverrini*: potential target for DNA-based diagnosis. Infect. Genet. Evol. 21, 443–451.

Tinga, N., De, N., Vien, H.V., Chau, L., Toan, N.D., Kager, P.A., Vries, P.J., 1999. Little effect of praziquantel or artemisinin on clonorchiasis in Northern Vietnam. A pilot study. Trop. Med. Int. Health 4, 814–818.

Touch, S., Komalamisra, C., Radomyos, P., Waikagul, J., 2009. Discovery of *Opisthorchis viverrini* metacercariae in freshwater fish in southern Cambodia. Acta Trop. 111, 108–113.

Traub, R.J., Macaranas, J., Mungthin, M., Leelayoova, S., Cribb, T., Murrell, K.D., Thompson, R.C., 2009. A new PCR-based approach indicates the range of *Clonorchis sinensis* now extends to Central Thailand. PLoS Neglected Trop. Dis. 3, e367.

Umesha, K.R., Kumar, S., Parvathi, A., Duenngai, K., Sithithaworn, P., Karunasagar, I., Karunasagar, I., 2008. *Opisthorchis viverrini*: detection by polymerase chain reaction (PCR) in human stool samples. Exp. Parasitol. 120, 353–356.

Viyanant, V., Brockelman, W.Y., Lee, P., Ardsungnoen, S., Upatham, E.S., 1983. A comparison of a modified quick-Kato technique and the Stoll dilution method for field examination for *Opisthorchis viverrini* eggs. J. Helminthol. 57, 191–195.

Vo, T.D., Waikagul, J., Thanh, B.N., Vo, D.T., Nguyen, D.N., Murrell, K.D., 2014. Endemicity of *Opisthorchis viverrini* liver flukes, Vietnam, 2011-2012. Emerg. Infect. Dis. 20, 152–154.

Waikagul, J., Dekumyoy, P., Chaichana, K., Thairungroje Anantapruti, M., Komalamisra, C., Kitikoon, V., 2002. Serodiagnosis of human opisthorchiasis using cocktail and electroeluted *Bithynia* snail antigens. Parasitol. Int. 51, 237–247.

Watthanakulpanich, D., Waikagul, J., Anantaphruti, M.T., Dekumyoy, P., 1997. Evaluation of *Bithynia funiculata* snail antigens by ELISA-serodiagnosis of human opisthorchiasis. Southeast Asian J. Trop. Med. Publ. Health 28, 593–598.

Watwiengkam, N., Sithithaworn, J., Duenngai, K., Sripa, B., Laha, T., Johansen, M.V., Sithithaworn, P., 2013. Improved performance and quantitative detection of copro-antigens by a monoclonal antibody based ELISA to diagnose human opisthorchiasis. Acta Trop. 128, 659–665.

WHO, 2006. Preventive Chemotherapy in Human Helminthiasis: Coordinated Use of Anthelminthic Drugs in Control Interventions: A Manual for Health Professionals and Programme Managers. http://whqlibdoc.who.int/publications/2006/9241547103_eng.pdf.

WHO, 2011. Report of the WHO Expert Consultation on Foodborne Trematode Infections and Taeniasis/Cysticercosis. World Health Organization, Geneva Switzerland, 59 pp.

Wongratanacheewin, S., Sirisinha, S., 1986. Analysis of *Opisthorchis viverrini* antigens by redioimmunoprecipitation. Asian Pac. J. Allerg. Immunol. 4, 62.

Wongratanacheewin, S., Chawengkirttikul, R., Bunnag, D., Sirisinha, S., 1988. Analysis of *Opisthorchis viverrini* antigens by immunoprecipitation and polyacrylamide gel electrophoresis. Parasitology 96, 119–128.

Wongratanacheewin, S., Pumidonming, W., Sermswan, R.W., Maleewong, W., 2001. Development of a PCR-based method for the detection of *Opisthorchis viverrini* in experimentally infected hamsters. Parasitology 122, 175–180.

Wongratanacheewin, S., Pumidonming, W., Sermswan, R.W., Pipitgool, V., Maleewong, W., 2002. Detection of *Opisthorchis viverrini* in human stool specimens by PCR. J. Clin. Microbiol. 40, 3879–3880.

Wongratanacheewin, S., Sermswan, R.W., Sirisinha, S., 2003. Immunology and molecular biology of *Opisthorchis viverrini* infection. Acta Trop. 88, 195–207.

Wongsaroj, T., Sakolvaree, Y., Chaicumpa, W., Maleewong, W., Kitikoon, V., Tapchaisri, P., Chongsa-nguan, M., Cross, J.H., 2001. Affinity purified oval antigen for diagnosis of *Opisthorchiasis viverrini*. Asian Pac. J. Allergy Immunol. 19, 245–258.

Wongsawad, C., Wongsawad, P., 2012. *Opisthorchis viverrini* and *Haplorchis taichui*: development of a multiplex PCR assay for their detection and differentiation using specific primers derived from HAT-RAPD. Exp. Parasitol. 132, 237–242.

Wonkchalee, N., Boonmars, T., Laummaunwai, P., Aromdee, C., Hahnvajanawong, C., Wu, Z., Sriraj, P., Aukkanimart, R., Chamgramol, Y., Pairojkul, C., Juasook, A., Sudsarn, P., 2013. A combination of praziquantel and the traditional medicinal plant Thunbergia laurifolia on *Opisthorchis viverrini* infection and cholangiocarcinoma in a hamster model. Parasitol. Res. 112, 4211–4219.

Worasith, C., Kamamia, C., Yakovleva, A., Duenngai, K., Wangboon, C., Sithithaworn, J., Watwiengkam, N., Namwat, N., Techasen, A., Loilome, W., Yongvanit, P., Loukas, A., Sithithaworn, P., Bethony, J.M., 2015. Advances in the diagnosis of human opisthorchiasis: development of *Opisthorchis viverrini* antigen detection in urine. PLoS Neglected Trop. Dis. 9, e0004157.

WPRO, 2017. Expert Consultation to Accelerate Control of Foodborne Trematode Infections, Taeniasis and Cysticercosis. World Health Organization Regional Office for Western Pacific, Manila Philippines, 16 pp.

Xiao, S.H., Hui-Ming, W., Tanner, M., Utzinger, J., Chong, W., 2005. Tribendimidine: a promising, safe and broad-spectrum anthelmintic agent from China. Acta Trop. 94, 1–14.

Xiao, S.H., Utzinger, J., Tanner, M., Keiser, J., Xue, J., 2013. Advances with the Chinese anthelminthic drug tribendimidine in clinical trials and laboratory investigations. Acta Trop. 126, 115–126.

Kolibri 360 GmbH
Ferdinand-Jühlke-Straße 7
99095 Erfurt, Deutschland
produktsicherheit@kolibri360.de